"Nadzam's prose is just gorgeous… A reader can swim in these sentences and soak up the landscape via the prose with great pleasure."
—Aimee Bender on Bonnie Nadzam's *Lamb*

"I started reading [Jamieson's prose] and couldn't stop… Jamieson, by elucidating our past failures and casting doubt on whether we'll ever do any better, situates it within a humanely scaled context."
—Jonathan Franzen on Dale Jamieson's *Reason in a Dark Time*

LOVE IN THE ANTHROPOCENE

DALE JAMIESON

BONNIE NADZAM

OR Books

New York · London

© 2015 Dale Jamieson and Bonnie Nadzam

Published by OR Books, New York and London
Visit our website at www.orbooks.com

All rights information: rights@orbooks.com

First printing 2015

Cataloging-in-Publication data is available from the Library of Congress.
A catalog record for this book is available from the British Library.

ISBN 978-1-939293-90-9 paperback
ISBN 978-1-939293-91-6 e-book

Text design by Bathcat Ltd. Typeset by AarkMany Media, Chennai, India.
Printed by BookMobile in the United States and CPI Books Ltd. in the
United Kingdom. The U.S. printed edition of this book comes on Forest
Stewardship Council-certified, 100% recycled paper.

FOR MEI AND JEREMY

"If we don't change direction soon,
we'll end up where we're going."
—*Professor Irwin Corey*

CONTENTS

INTRODUCTION:
THE ANTHROPOCENE

How and where will love arise in a world in which nature has become almost entirely an artifact? We invite you to imagine human beings living in worlds of the future that have been remade and are almost entirely managed by human action: rivers, lakes, oceans, forests and fields are as meticulously planned and technologically maintained by humans as are cities and their systems of transportation and utilities. The weather is profoundly affected by the inadvertent consequences of human action, and by clumsy attempts to correct these perturbations and bring them under intentional human control. The other animals with whom we share the Earth survive only at the margins, and only if their continued existence can be justified to planetary managers.

The world we inhabit now is not really so far from this imagined world. Wild elephants and rhinos are now being protected from poachers by unmanned drones while "Gorilla docs" intensely monitor the world's last wild population of Mountain Gorillas, staging medical interventions when necessary. An integrated system of mobile gates that will protect the history and treasures of the city of Venice by isolating the Venetian Lagoon from the Adriatic Sea is now 80 percent complete. Should you find yourself in Dubai you can downhill ski in the Mall of the Emirates, even on the hottest summer days.

The fact is that virtually all of the terrestrial biosphere has been transformed by human action, and the oceans are not far behind. More seafood that people consume is produced by aquaculture (fishfarming) than by fishing. Twenty million tons of human trash is in the world's oceans, much of it concentrated in a several hundred thousand square mile area known as the "Great Pacific Garbage Patch." Millions of pieces of debris from satellites, rocket boosters, and lost equipment are now orbiting the planet.

The few places that are largely immune to human impacts are found on mountaintops, deserts, and in polar regions—but now even these areas are being transformed due to climate change. Fossil fuel–driven climate change is opening up new regions of the arctic to oil and gas exploration, which may lead to more fossil fuel–driven climate change, which may lead to more areas of the Earth being exploited for fossil fuel production, and on and on it goes.

The human transformation of the planet is fed by technology that continues to develop according to Moore's Law (roughly that computing power doubles every two years), though the reliability of new technologies is often "iffy" and new threats emerge daily as we become increasingly dependent on electrons for maintaining our financial records, preserving our cultural history, protecting our security, and defining who we are and how we present ourselves to each other.

What we ask you to imagine is not a world of fantastic, mind-blowing innovations in technology and radical alterations in human behavior, or an apocalyptic wasteland in which all of the worst of doomsayers' predictions have come to fruition. Rather, we invite you to consider with us the path we are already on and where it might lead.

We explore our questions—about love in a world in which nature has become almost entirely an artifact, and what this might mean for human loving relationships—by telling stories and sharing meditations, not by issuing predictive declarations that are supposed to provide answers. Prediction is always risky, but especially so when human action is involved. There is little about the world of the future that is fixed and determinate. As the good Professor Corey insinuates in the epigraph, awareness of where we are going can lead us to change course.

A good example of this concerns chlorofluorocarbons, which were marketed under such brand names as Freon. For most of the twentieth century, we used these chemicals as refrigerants, solvents, and propellants. Had brilliant scientists not alerted us to the consequences of using air conditioners and spray cans that used these chemicals, the ozone layer would have eventually been destroyed and the Earth rendered a lifeless planet.

This same example illustrates how forces driving change can interact in surprising ways beyond anyone's intentional control. Early twentieth century chemistry combined with mass consumption caused the problem of ozone depletion. More advanced chemistry combined with remote sensing and a brief era of international cooperation allowed us to solve the problem before substantial damage had been done.

While there are many reasons to be concerned about the future, how bad or good it will be will depend to a great extent on the values of those who will live in that world. Consider, for example, present-day New York City. "I ♥ NY" say many of today's residents and tourists. But the seventeenth-century inhabitants of Manhattan, the Lenape Indians, would have probably been appalled by much of what present-day New Yorkers and visitors admire: skyscrapers, subways, museums, designer brands, bars and restaurants with a buzz. For all we know, we stand to future people in the way that the Lenape stand to us. We may be appalled by their tastes and desires. They may despise what we think of as the treasures that we sacrifice to bequeath them. Knowledge changes, but so do preferences and desires.

What, you may wonder, does this all have to do with love? Don't we love our New York children as the Lenape loved their Manhatta children? What do managed, manicured parks, the incidence of skyscrapers, geoengineered climate, nanoparticles and the increasing use and dependence on virtual reality have to do with love? One answer is wait and see. But to anticipate what we hope you will see, think for a moment about how inseparable love is from nature in our actual experience—whether it involves those special places where we have fallen in love, activities we love to engage in, or even the love of nature itself. More profoundly, where does nature end, and we begin? From the point of view of biogeochemistry, nature is the carbon cycle, the nitrogen cycle, and other natural cycles. We, like objects on this planet, are living embodiments of these cycles. Our breathing and respiration are instances of the same cycles that govern the atmosphere; our circulatory system as well as various cellular processes are instances of the hydrological cycle; digestion and metabolism recapitulate the soil cycle; and we are as subject to the laws of thermodynamics as any planet or star. To neglect the natural world from which we are constituted is to ignore the very matter of our own bodies—the hands with which a daughter caresses the face of her dying father, the rapid, heated breath of a passionate kiss, the arms in which we cradle a child or a beloved pet. It is also to ignore the hands we stuff in our pockets when we look away from suffering, the arms we cross over our chests to defend ourselves from perceived threats, and the hardened expressions on our faces when we stare down our enemies. Our bodies—how we love with them and use them to navigate a first date to the movies, to waltz across a polished floor, to give up a subway seat to an elderly woman—are inseparable from nature.

Today, the main driver of change on planet Earth is not volcanic activity, shifts in tectonic plates or variation in solar radiation, but the growing human population and its demand for energy, food, information, services, and the need to dispose of waste products. The term "Anthropocene" was coined in order to mark the scale and significance of such human impacts on the planet. In the last 250 years, humans have caused not only climate change, but also species extinctions, desertification, ocean acidification, ozone depletion, pollution and more besides. First used by the biologist Eugene Stoermer in the 1980s, this term "Anthropocene" came to widespread public attention in 2000 when he co-wrote a short article with the Nobel Prize-winning chemist, Paul Crutzen, suggesting that we may be entering a new geological epoch distinguished by these widespread and profound human impacts.

The Holocene, as the current geological epoch is officially known, began 11,700 years ago. During this time Earth's systems—its biosphere, atmosphere, lithosphere, and hydrosphere—have been unusually stable. Biologically modern humans emerged about 200,000 years ago during the Late Pleistocene, but virtually everything we associate with humanity (e.g., agriculture, cities, writing) developed during the last few thousand years. At the beginning of the Holocene there were probably about six million people living as hunter-gatherers. Today there are more than seven billion people, most living in highly complex urban societies.

While the word "Anthropocene" is new, the idea has been around since the nineteenth century as scientists, theologians, and naturalists struggled to give voice to the dawning realization that humanity, with its technological power, was remaking the planet. In 1907, the French philosopher Henri Bergson wrote:

> *In thousands of years...our wars and revolutions will count for little...but the steam engine, and the procession of inventions of every kind that accompanied it, will perhaps be spoken of as we speak of the bronze or of the chipped stone of pre-historic times: it will serve to define an age.*

Even earlier, in 1864, the American polymath George Perkins Marsh identified human agency as responsible for large-scale changes in the planet, some breathtaking and many worrisome. He was struck by the massive changes he witnessed from the time he was a child in Vermont living alongside Native Americans, to the deforestation and desertification he saw as a diplomat working in the Mediterranean region.

Many of us have witnessed changes as dramatic as those Marsh observed—if not directly in our own neighborhoods, then certainly in distant cities and countries, through various media, including television news, documentary films, and so on.

With change so constant, present, and immediate, it is not surprising that the idea of the Anthropocene has fallen on fertile soil. Over the last decade, websites, organizations, scholarly books, novels, a refereed scientific journal and even "creativity" workshops have been established that are devoted to the study of the Anthropocene. A proposal to declare the Anthropocene a new epoch in Earth's history is under formal review by the International Commission on Stratigraphy (ICS), the authoritative scientific body that makes decisions about the geologic time scale.

Geologists write Earth's history on the basis of what is found in layers of the Earth's crust. For example, the transition that occurred about sixty-six million years ago between the Cretaceous and the Paleogene is marked by a thin layer of iridium-rich sediment and a break in the fossil record (dinosaurs before, no dinosaurs after—an asteroid strike may have delivered the iridium and caused the extinctions). Perhaps the ICS will view as the mark of the Anthropocene the radioactive isotopes that began to be laid down in the Earth's crust in the 1940s as a consequence of nuclear testing. It may be more difficult to find in the stratigraphic record distinctive traces of the systematic human transformation of the lithosphere, atmosphere, hydrosphere and biosphere that have occurred as a result of industrialization and mass consumption. There is no guarantee that a record of who we are, how we lived, and in what we found meaning, will be encoded in the Earth's crust.

Whatever its fate as a geological category, the Anthropocene is an important concept for understanding how we live and what the future may hold. Our current cultural and moral world differs markedly from that of our forebears. These differences are likely to increase exponentially in the next few centuries as we continue to seek immortality, and renew our quest to restore or even improve the natural world whose order we have ruptured.

Technology and the Anthropocene are joined at the hip. The humanity that has transformed nature is now organized in highly complex systems bound together by air travel, oil and gas pipelines, electrical wires, highways, train tracks, fiber optic cables, and satellite connections; this enables "action at a distance" that would once have seemed inconceivable, whether as a conversation on Facetime or through the instantaneous transfer of wealth and resources from one side of the world to another. These possibilities affect the nature of our relationships and our conception of agency. We now feel empowered to save a child in a faraway village by

clicking on a link or making a phone call and pledg-
ing a contribution. We can adopt children from
cultures we never knew existed and carry on love
affairs with people who not long ago would have
been faceless and mysterious. Yet at the heart of this
power is a sense of helplessness when it comes to
stemming climate change and other transformations
that herald the arrival of the Anthropocene. While
together we are bringing about these changes, the
sense of agency and responsibility that was central
to the Enlightenment tradition that gave us so much
of our literature, politics, and morality, seems to be
disappearing in the high-technology, mass society
that we have created.

In a world of greater resources and fewer
well-insulated spheres of life, everything becomes
increasingly fungible, from wombs to kidneys to
artworks to lives themselves. Trade-offs between the
near and dear, and the remote and strange, become
possible in a way they never were before. Sitting in
his rooms in Cambridge, the late nineteenth century

philosopher Henry Sidgwick could read about famine in Bengal, but there was little he could do about it in real time except to decry it. You, on the other hand, can immediately empty your bank account and make something happen on the ground in a disaster zone—maybe save some lives, maybe increase the opportunities for corruption. Or instead you could go in for some online gambling on your children's future. Or you could just leave everything to your canary. Everything seems possible but nothing seems certain or to matter. In a world where everything affects everything else and no one feels decisive over much at all, the distinction between causation and complicity has become fraught. When I drive my SUV to the 7-Eleven for a Slurpee am I causing climate change, contributing to it, complicit in it, or does it not matter what I do? What, if anything, turns on these distinctions? Is San Diego Gas and Electric giving us what we want, or manipulating us because we have no choice but to pay our bills? The list could go on. And on and on.

Such questions and their lack of answers can lead to a crisis in meaning. Human life has traditionally been lived against the background of a nature that is seen as largely independent of human action. The Book of Matthew tells us that the sun shines on the just and unjust alike. Once geo-engineering is perfected we may be able to fix this oversight of nature. But what becomes of the message of humility and compassion that this teaching evokes?

Many questions remain but this much we know: the Earth is rapidly changing, and humanity is a prime mover. What we are losing is substantial and the Anthropocene will in many ways be a diminished world. But will it be a world in which our most important values shrivel? Will it provide a platform for developing new rituals, ceremonies, relationships and rites of passage that are meaningful and appropriate to this new epoch?

The Anthropocene presents us with many questions. Our focus is on how and where love—as we know it today, and as our forebears understood it—can arise in the world of the Anthropocene. This is not a narrow scientific question. It contains a fundamental human challenge. The story of the Anthropocene begins with geology, but is ultimately a story of the human heart.

Finally, a word on our collaboration. We are each, and together, responsible for every word in this book. The authorial voice is neither Dale Jamieson's nor Bonnie Nadzam's, but a third voice which we have jointly created.

DALE JAMIESON, *New York City*
BONNIE NADZAM, *Fort Collins, Colorado*
June 2015

ONE: FLYFISHING

"Fishing!" the old man cried. He had thick white hair that curled around his ears, and hands that trembled as he raised them in the air. "Who's fishing?"

"Dad," said the woman beside him, in a tone meant to make him lower his voice.

But the old man did not lower his voice. In preparation for an announcement to everyone on the maglev now speeding its way up the plains and into the foothills toward Wild Rivers, he straightened his neck and lifted his chin. "You let me know on our way down," he hollered, "if anybody you saw did any real fishing."

The woman sighed. "How about some lunch?"

▼

Across the aisle, the tall, dark-haired man and his young daughter sat side by side, she with her arms crossed, chin tucked, he scanning the world outside the window. He pointed and then reached back to jostle her arm. "See that?"

The girl kept her gaze fixed on the floor.

"Wow," the man said. "God. When your mother and I came up here it was just dirt and rocks. None of this was here. No trees, none of those hedges. Not much river to speak of at all." He shielded his eyes with his hand and looked out into the daylight. "I mean you'd hardly know it was the same country."

"We could have gone fly-fishing right from our living room. I don't see why we had to come all the way out to the middle of nowhere."

"Oh come on. It's not the same. And there'd be no real fish to eat if we did it at home."

"I am not eating fish." She looked sideways at her father, then glanced back the other way and caught the eye of the woman across the aisle who it seemed had overheard the exchange.

"She doesn't want to fish," the man explained to the woman across the aisle, and shrugged. He smiled at the woman, and at the old man beside her.

The old man laughed. "That's right honey," the old man said. "You got to know when to count your losses. Isn't that right?" He looked at the girl. "We're the only idiots on the train I guess."

The train sped through the canyon and into higher country, increasingly bright, green, and, off in the distance, white with snow.

"Beautiful," the dark-haired man told his daughter. He pointed again.

"It's not real, Dad."

"What are you talking about? Real grass. Real trees. All growing in the rain and sunshine."

"Why do we have to stay for five days?"

He ignored her question. "And since when did things not being real bother you?"

"This is different. We're pretending it is."

"You know, some people—a lot of people—would much rather spend a few days in the mountains with their loving father than another long weekend in town with their friends."

"Name one."

"Will you try to enjoy yourself? For one day? This is important to me." He was quiet a minute, and the train began to slow around a curve. "If it's that awful we can turn around and come back."

"I thought you couldn't get your money back."

"I can't."

"One day?"

"On one condition. First, uncurl your lip so I know you're listening."

She set her hands in her lap and turned a placid face to her father.

"If you start getting shitty, the deal's off, and we're there all five days."

"You want to torture yourself?"

"No," he said. "Just you. I'm going to have a great time no matter what."

They sat in silence. The old man and woman beside them were finishing their lunch. She tried to help him wrap the uneaten remains of a sandwich but he pushed her hands away.

"We talked about this," the woman said in a low voice.

The old man leaned over to look at the young girl, his eyes full of light but his face stern. "Don't let them bullshit you."

The girl laughed. "I never do."

"No rocks," he said, shaking his head. He raised his voice and hollered down the train. "No rocks!" Several passengers turned back and looked.

"Oh for godsakes, Dad."

He raised his hands as if he were a preacher holding forth. "Nobody falls, nobody drowns." He surveyed his fellow train passengers, expecting amazement. Realizing he had no audience, he looked again at the girl. "Is that a river? That's what

I'm asking you." The girl stared at him. He waved his hand, dismissing all of it and turned his face to the window.

"Well," the woman said, smiling across the aisle at the man and his daughter. "We won't miss the mosquitoes."

The girl's father laughed. "Or days with no fish."

"Oh dear God. Not for a thousand dollars a day."

"I hear you."

"Ever fish for free?"

"I've heard there are places in northern Europe. But getting there is hardly free."

"I can imagine."

"I went a couple times when I was little," he volunteered. "In Michigan."

"No kidding. Was it still State Park?"

"Must have been."

"They say the fish doesn't taste the same."

"I've heard that."

"You don't remember any difference?"

"Nah."

"How many days will you be up here?"

"We're still negotiating," he said, jerking his head in the direction of the girl, who sat slumped with her arms still crossed.

The woman looked at the girl. "Ever had fish?"

She shook her head.

"You're in for a treat."

▼

As they climbed higher, the man told his daughter about night fishing on the Pere Marquette River, years ago. "I was just a little younger than you are now, twelve or thirteen. Your Uncle Will was sixteen."

"Grandpa took you?"

"My grandpa took us. Your grandpa was working."

"Oh."

"We went in early summer. It was a half a day's drive. Trout always come up to feed when the light is changing—you'll see in the morning. In Michigan the sunset didn't happen until almost ten thirty, so we were sometimes fishing really late, into the pitch black hours."

He told her about a night when the pine trees were like black cut-out lace against the iridescent, navy blue sky. There was a big moon, and he and his grandfather and brother were giving it one more go.

"Must have been about ten, ten-thirty at night. Will had a big one on the line, and he just let out line and let out line until it was a quarter mile down the river, and he was racing through the woods, jumping fallen logs to keep up with it."

The girl was listening.

"I mean there was a moon but it was pretty dark in the trees, and he was racing along the river with that fish. God."

"Did he get it?"

"He did."

"And you ate it?"

"We threw it back. We almost always threw it back."

"To save the fish?"

He looked at her. "That's right."

"But you must have eaten them sometimes."

"You bet. Best way is to wrap them in foil with a stick of butter."

"Oh my God. A whole stick?"

"A whole stick. Real butter. Salt and pepper. Slice an onion. Seal it all up and put it on slow burning coals."

"In a real fire?"

"A real fire."

"Will we have a real one?"

The man tipped his head, considering.

"I didn't think so," she said.

"But maybe," he said. "We'll see."

"Do you even know how to build one?"

"I think we could figure it out."

"What if we got lost and really needed one? Like to signal someone like the Indians did in ancient times?"

Was she being ironic? "Sweetie," he said, and opened his hand on the crown of her head. Her hair was silky and fine. "No one gets lost."

By the time they pulled into the park it was dark, and late. They found their rental cabin and he unpacked his gear on the couch, she on the bed with the six fluffy pillows.

▼

In the morning he made coffee and toast in the cabin and they went out to the river.

"Big sky," he said. "That's what they call it."

"I thought the mountains would be bigger."

"Well, we're already pretty high up."

All around them the willows and tamarack were green. Small placards indicated the names of surrounding grasses and flowers: lady's tresses and hollyhocks.

"Actually be kind of exciting to find a weed," he said. "That'd be news."

"What's the matter?" she said. "Don't you find all the plants real and natural?"

He looked at her. "That wasn't sarcasm was it? That'd be a deal breaker."

She batted her eyelashes at him and he laughed. He gave her a license and park pass. Within minutes the ranger approached and distributed the flies they'd need for this season.

"Brought your own rod," he said. He was a young guy with bright straight teeth and a trimmed brown goatee, and wore a shirt stitched over the breast pocket with the Wild Rivers, Inc. logo. "That's a beauty," he said, looking at the rod.

"My grandfather's," the man said, turning it in the daylight before them.

"No kidding. May I?" The ranger took the rod and reel, weighed it, and drew his arm back as if to cast. "He took good care of it," he said.

"Well. He wanted me to be able to fish with it."

The man and the ranger exchanged an odd smile, and the ranger turned to the girl. "You guys sharing or you want your own?"

"I don't want one," she said.

"Give her one of her own."

The ranger took a black canvas tube from his gear and gave it to the girl. "Your dad can help you set it up." He handed a small electronic device to the man; it was an inch square and attached to a clip that would hang from a shirt or a pocket. "And here's our little pamphlet," the ranger said, handing one over. "Regulations and guidelines. Little bit of history."

On the inside flap there was a black and white photograph of the river from 1937, and beside it, a photograph of the park now. The images were identical in rock formation, tree line, and vegetation.

"It's remarkable," the man said.

"Only difference is," the ranger said, "what we got now is permanent. And no bears."

"Can't argue with that."

"Tell you what," the ranger said, looking at the river. "Suicidal fish. That's what they are. You're in for a good day."

The man handed his daughter the electronic device. She turned it over in her hand. "Just a tracking device," the ranger said. "If you need one of us, use that. Otherwise we'll be out of sight, out of mind."

The girl raised an eyebrow at the ranger.

"Leave it off if you prefer," he said, and he left them on the river.

▼

The man and the girl stood at the edge of the water and he talked her through threading the eyes of the rod and setting the reel in place.

"Having an awful time?" he asked.

"If I say yes, the deal's off and we're here for five days."

He stepped out onto the riverbed and looked back at her. "No rocks!" he cried in a strained, old man voice. She doubled over with laughter. He dragged his foot back and forth along the riverbed. "It's like a running track," he said.

"No one falls," she said, imitating the voice.

"No one drowns!" He motioned with his arm for her to join him, and she stepped into the water in her boots. "Do you remember how to do this?"

"Show me again."

"You want to let your line out like this, about three times the length of your rod."

"Ok." She followed suit.

"Hold the rod right here," he moved her hand, "like this." He showed her on his own. "Like you're shaking someone's hand. Good. Now, pull it back to about 10 o'clock." She looked at him quizzically, as he showed her where 10 o'clock would be. "Not your arm, just bend at your elbow, and then, like this." He showed her. She copied him. "Not bad, not bad."

They practiced false casting a dozen times and the sun rose up overhead. He caught a fish almost immediately. "Sure makes you feel like a pro."

"Uh-oh," she said, grinning. "Was that a shitty statement? Are you wishing it was harder to catch one? Like in the olden days?"

"The olden days?" He looked back at her shaking his head but smiling. "Tell you what, sweetheart. You'll know a shitty statement out of me when you hear one." He wet his hands in the river water to slime the scales, held the fish firmly in one hand, removed the hook, and returned it to the water. It seemed like a small fish.

"Don't let a ranger see us do that," he said.

"Would they kick us out?"

"There's a fine."

"Why?"

"People don't like it if their fish has already been caught. It doesn't feel clean."

"Oh."

They'd paid for the minimum: space on both banks, and four hundred meters up and down river. They could see the daylight flashing on the lines and equipment of other park visitors. The sky was a hard, empty blue. Except for the sound of water, it was dead silent. On the north side of the river, half a mile out, a swath of company lodgepole pine.

"Can you smell it?"

"What is it?" She asked as she scrunched up her nose.

"Those trees," he said, nodding at them. "What a fucking blessing."

"Dad!"

"Well you should've seen them fifty years ago. What was left of them, I mean."

"I've seen pictures."

"I guess you have."

"Alan Jefferson trees," she said.

"Alan Jefferson?"

"Duh, the man who invented them?"

"Shows you how much I know."

"Not too fucking bright, Dad." He whirled around and they looked at each other and laughed. He pointed at her.

"Stop it!" he said, but he was laughing.

In the next hour she caught her hook and line in the willow behind them several times, and lost a couple of flies in a stand of alder. He showed her again, and then again, how to re-tie a fly, how to make the knot, and how to feed it properly through all the eyes down the length of the rod.

"If we were out here in the 'olden days,' we'd have to determine which flies to tie on there based on the season and the kind of fish."

"How would we know that?"

"Well we'd pay attention to what would've been hatching."

"What would've been hatching?"

He laughed. "I have no idea."

She looked up at the sun. "It's really hot."

"How about some sunblock and a snack?"

The park would have provided refreshments but he'd wanted it to be for her as it'd been for him as a boy. They had sandwiches with cheese and pickles and thick slices of Vidalia onion, and coffee out of a thermos, even in the heat of day.

"You had coffee when you were that little?"

"Sometimes a little cold beer too, to put fat on my bones."

"Can I try some?"

He looked at her.

"Come on. I know Uncle Will was giving you beer when you were fourteen."

"One," he said. "One beer."

"I'll stay a second day for that."

"Did I ever tell you," he said, leaning back on one hand and crossing his ankles, "about the por-cupine?" He took a bite of his sandwich.

She opened her sandwich, peered at the onion and pickle, and closed it again. "No."

"My grandfather beat a porcupine to death with a shovel."

She stared at him, eyes wide.

"It was pretty horrible," he agreed.

"Why did he do it?"

"It was a pest, he said."

"What happened?"

"It was one of those fishing nights, and he just took out after this porcupine that had come up to the porch. Smelled our garbage probably. I could hear the whole thing."

"Whoa."

"That was our last trip."

"Because of what he did?"

"Your Uncle Will pulled in trout after trout that day and our Grandpa didn't catch a thing. He was mad as hell, pacing the riverbank like the shore patrol. So then Grandpa takes off after the porcupine. Beats it to hell. 'Why'd you go and do a thing like that,' Will asked him. 'It's a creature. It wants to live.' 'It's a pest,' Grandpa said. And then we left the next day. Didn't talk the whole drive home. Will was pretty pissed off."

"So what. Your grandpa didn't want to eat any of that trout that Uncle Will caught?"

"Stubborn pride runs in the family."

She caught his eye and they grinned at each other.

Suddenly a shadow passed over the water and clouded the grass behind them. All at once the man pointed up into the sky and was on his feet. "Look at that!"

She got up and looked too. "It's huge!" she hollered. "What is it? A hawk? An owl?"

They stood watching it shrink and disappear into the vacant blue, then looked at each other, the tension of an unspoken question between them.

"They don't have holograms that can do that," she said. "Do they?"

"No way." At least he didn't think so. It must be some new kind of technology.

"It was huge! What was it?"

"I'm not exactly sure." He knew it was supposed to be an eagle, and that there were no real eagles. He looked at the back of his daughter's head, her hair shining the daylight. He felt a weight in his chest, and heat burning behind his eyes.

"It was an experience that's for sure," he said, "seeing that."

"Do you think anyone will believe me if I tell them?"

"What are you going to tell them?"

"I don't know."

▼

After lunch she got her first good cast of the day and hit right away. It was a golden, and it was beautiful. It had deep yellow flanks with red, horizontal bands along the lateral lines and about ten dark, vertical, parr marks on each side. Its dorsal, lateral and tail fins had white edges. It had to be just over ten inches long. Perfect.

"What do I do?" Her eyes were wide.

He laughed. "Reel it in, girl!"

"It feels caught on something."

"It is. Your hook. You have to really pull."

"I can't."

"Put some muscle in it. It's fighting you. On one knee," he said, "get a good advantage. Steady so you can pull it in just like that. That's right. Take the line right above it. See it? See it? God that's a beautiful fish. Even Grandpa would have said so."

She held the line up and admired the fish. It twisted and flopped on the line. Her small arm strained with the weight of it.

"Go ahead," he said. "Take it in your hand."

She did, twisting up her face and looking back and forth from her father to the fish.

"Now reach into its mouth there and get the hook. You can pull it right out. Go on."

"I can't get it."

"Yes you can. You have to hold the fish. Really hold it."

"Daddy." It was still flipping. She pulled and turned the hook gently in the fish's mouth. "I want to put it back in the water." Its gills were working hard. "It's looking at me."

"No it isn't. Here. You got to use a little pull, honey." He showed her. She pulled. She didn't think it would be so hard. It was like tearing the hook out through bone.

"I want to put it back."

He laughed. "Squeamish?" Her eyes glittered with tears. He shook his head, met her eyes. "This is part of it."

"I hate it."

"You have to remember this is what fish are made for. They're hatched and grown up there, and they slide down the river for people like us to catch."

"And eat?"

"And eat."

"But I don't want to eat it."

"It's what they're for. Think of all the good river days that fish had."

He reached out and she let him take it from her. In one swift movement he knocked its head against one of the vertical snub-nosed stones strategically placed along each side of the riverbank. She held her breath in her lungs, lips pulled tight into her mouth, and watched.

He looked up at her. "You touched it with dry hands so we couldn't have put it back."

She was crying openly now. "I can't look at it."

"You don't want to learn to clean it?"

"No."

"Not even with your new knife?"

She shook her head.

"Will you eat it?"

"I don't know."

"I'd be awfully sorry if you didn't get to try it once."

"Because it's so expensive," she said, running her sleeve beneath her nose.

"Aw, honey. That's not it."

She nodded, her face in her hands.

"It's different when you kill it yourself," he said. He put his hand on her shoulder. "Isn't it?" He reminded her that they'd paid for this already, and that her mother was looking forward to having some.

"I don't want to do this anymore Dad."

"Do you want to go home?"

"Can we just stop for a while?" she asked.

"We could walk the river some," he said. "It's very pretty." He looked at it. The bends that would never oxbow. The rills among the willow. He gestured for her to sit beside him on the ground. She drew her knees up into her chest and wrapped her arms around them.

"It's a perfect river," he said. "Isn't it?" His shoulders were a little slumped, their rods on the holding stones placed in the sand to keep the reels out of the dirt. He opened the thermos and drank some coffee. They sat in the quiet until she looked up at him.

"Do you really want me to try it, Dad?"

He smiled at her and jostled her knee "It's ok," he said. "I understand."

"I'll eat it," she said, "or at least, I'll try it. Promise."

"You don't have to."

"What if we made it how you used to?" She extended her skinny legs out in front of her. The sun was warm on her skin. "I mean, if there's a place to build a real fire." They were both looking down at their own legs, and out at the water.

"Be expensive, honey."

"If it's the only fish, though?"

He looked at her and grinned. "Stick of butter?"

"Stick of butter. And I'll chip in."

"Where's that little electronic thing. We can send for the ranger and find out."

She took it out of her back pocket and handed it to her father. "However much it is," she said, "I'll pay half."

"That's ok. This is my gift to you." He turned the device over in his hand and switched it on. He studied it a moment, pressed its small screen twice, and put it in his own pocket. "Want to cast a little more while we wait for him? It feels good on my old knees to stand in the river."

"Can we kick off our shoes?"

"Sure," he said, and stood up. "But let's get this pretty guy on ice first, so he'll last."

TWO: CARBON

The sky was as blue as a royal blue crayon and the sea glittered a mile from the street. Jack bought a hotdog from the kiosk and handed it to the homeless man they all called Rex for a plastic painted crown he'd worn for a season a few years back. The crown was gone but the name had stuck, and so had he: the man was always there, a permanent fixture, a bright-eyed, sunburnt and finely wrinkled face as familiar as the storefronts, high rises, and city park. He was never without his long brown coat, which he wore over a bare chest on days as hot as today, and he was always moving. Dancing. Singing. High-fiving little boys. Whistling, waving, talking in your face.

With a little bow he took the hotdog from Jack and raised it in the air, as if in a toast. "Thank you sir," he said.

Jack bought him a dog a couple three times a week. It wasn't exactly charity. It made Jack feel good, and Rex on his corner didn't seem so different from a lot of guys on bar stools at the corner pub. You could talk him up about last Saturday night's game. He'd tell you the best place on the shore for cheap street tacos, and the best palapa for greasy plantains and cold beer. And he was right. Course the guy had bad days too, temple to the concrete, sour stench of vomit and urine, face half hidden beneath his hood and nothing but streams of expletives for the men and women he'd usually greet with cheer. On such days the passersby who knew him might stoop down and slip a bill or two into his filthy coat pocket, but for the most part they walked by with their chins lifted, eyes pointed straight ahead. This was not the Rex they knew, and they obliged their acquaintance by politely ignoring him when he was in this state. Everyone knew how to do this; every block had its Rex.

It was late afternoon and Jack turned toward the glass doors of a steel and glass tower that disappeared into the brightness overhead. Behind him Rex sang out to a beautiful young woman in a coral colored dress. He had a nice voice, deep and even, and everyone liked to hear him sing. Sometimes he'd imitate female artists, trying out a high-pitched falsetto. It was a real attention-grabber. Now he danced a little imaginary waltz with the hotdog in the air. Jack smiled at him from the glass elevator and watched the familiar figure shrink and disappear as he floated up ten, twenty, fifty, one hundred stories.

▼

On the top floor there was an upscale restaurant and bar where he was to meet Ed. He took a seat near the glass wall and leaned his head against the window, looking out at the water. He glanced at his watch. He watched the ant lines of people along the parkway and the silver stripe of the rail glinting in the sunlight. He bit his nails. He took a deep breath, filled his chest and belly, and let it all out slowly. He ordered a second drink. Halfway through it a tall, lean man, with thick, light-streaked hair and resplendent in a beautiful dark suit and sunglasses, appeared in the window reflection behind him. Jack spun around.

"Oh my God." Jack's heart knocked hard in his chest. "It's you."

"It's me." Ed smiled as he opened his suit jacket with his thumbs and spun around. So easy, so cool. "You promised me real meat. So here I am."

Jack stood and the men hugged. "That I did."

Ed raised an index finger. "And no meal worms. So sick of meal worms."

"No, God." Jack grinned. "You look great." He ran his fingers back through his hair. "Was it a good interview?"

"I killed them."

"You always kill them."

"And never take less than twice what I deserve, when they come groveling."

"So? You here for good?"

"For good or for ill—if they pass the audition." The men sat down at the small table by the window. "Seriously though, the Persian Gulf project is winding down. The locals can handle it from here and your old friend needs something new. I seem to put myself out of work with every job I take."

"That's the idea, isn't it?"

"That's the idea."

Jack grasped Ed by the upper arm and jostled it. "It'll be just like when we were kids on the shore." The thought of it made a sweat break out beneath his arms at the small of his back. Jack and Ed. Ed and Jack.

"Easy," Ed laughed, and leaned back and crossed his legs. "What are you drinking?" He signaled the waitress and ordered a bourbon, neat.

Jack couldn't help it. "Do you need a place to stay? Really? Mi casa su casa. As long as you need."

"Down, boy." Ed pulled his chair out a little, away from Jack. "Let me look at you."

Jack turned his torso to the left, to the right. Stroked the dark stubble of his five o'clock shadow. "How will I do?"

"Tell me something," Ed said, turning away from Jack and looking out at the day. "Is it always this hot here?"

"Hotter."

Ed shook his head. "Thought I might get a break. Every job is someplace worse than the last."

"I bet it's been hell over there."

"Yeah, but the partying's crazy. Adversity focuses the mind and loosens the pants."

Jack nodded and looked into his drink.

"Oh come on, Jackie, don't be such an old lady."

"Don't call me Jackie."

Ed laughed. "The thing is if I take the offer, it's going to be seriously long term." He scrunched up his shoulders and shook out his arms and glanced around. "Makes me restless just thinking about it."

Jack leaned back, away from his friend, though everything in his body resisted the movement. "Tell me more." He sipped at his drink. "Something different about this one?"

"In the first place, you can't believe what this place needs. Without some serious work you guys are up to your ears in water in no time. We've got to build islands where there isn't any land, and wetlands where there isn't any water. Reefs, channels, locks, you name it. We're talking a whole new geography out there."

"Ten year project?"

"At least. Maybe never-ending. Like painting the Golden Gate Bridge."

Jack grunted. "Sounds familiar. And in the second place?"

"This is the best part. You guys have some serious money here with some real political clout behind it. We can up our game and do something really cool."

"Example?"

Ed explained. Most of these shoreline reconstructions are just that—trying to put back something that's been lost. Replace the functionality. Nobody cares much about how ugly it is and nobody's thinking co-benefits. Just put in the dikes and drains, get the job done, cash the checks, and go on to the next place. No innovation. Clumsy and short-term.

"You have better ideas," Jack said. He drained his glass and swirled the ice.

"What we're talking about here," Ed said, "is designing the shoreline that God would've designed if He were smart enough to know we were going to screw up the planet. We're going to start from scratch and instead of asking how to protect people and property from rising sea levels, we're going to ask what's the best possible shoreline we can imagine. That's what we're going to do if I take the job anyway." He smiled.

"So, 'co-benefits'"?

"Like on this project I just finished, these ten miles of buttes almost make a ring," he gestured with his hand, "and they're covered with running and hiking trails."

"People run in that heat?"

"I fucking don't. But in the morning other people do. And what happens is the trail design and artificial rock walls—it's like a labyrinth they run through, right?—it actually absorbs their body heat, which gets stored in underground batteries. Over time—an entire season, say—that body heat accumulates enough energy to run the pumps that drain the overflow when it's needed. Nobody's got anything on this stuff Jackie. I'm right on the edge."

"All this technology," Jack said, and whistled. He tried to signal the waitress, who missed his eye.

Ed leaned back and exhaled heavily. "All this technology."

"So running trails on buttes and everybody's farts up there will run the lights in their apartments. What else?"

"Much more. So much we can do, Jackie. This is the job everyone wants." He nudged his drink in the direction of his friend. "What about you? Same old same old?"

It sort of was.

The fossil fuel industry was so successful back in the day that running it in reverse was going to take a long time; there was enough carbon in the atmosphere to get him through to retirement.

"It's an ill wind that blows nobody good, as we say in the carbon removal business." Ed laughed at that, and Jack smiled, pleased.

But, Jack said, somehow he thought he would be working in a lab doing science. He raised his hand to signal the waitress, missed her again, and gave up. He flipped his napkin over, pushing his empty glass to the edge of the table. What he didn't imagine he'd be doing with his life, he told Ed, was managing people and remaking the atmosphere by remote control. There weren't many real scientists around anymore. Everyone was blue collar except the guy upstairs who walked in circles imagining new artificial tree designs, and the half-naked homeless guy dancing and singing out on the street.

"So what's wrong with being blue collar?" Ed asked. "What a snob you've become."

"I'm a snob? Look at that suit."

Ed smiled with all his teeth. They'd known each other long before suits. "It's a pretty nice suit isn't it?" He opened the lapel of his jacket again to show Jack the lining. "Look at that silk."

Jack leaned in and shook his head. "That's the most goddamned beautiful suit I've ever seen."

Ed closed his jacket. "I get what you're saying though, I do. Saving the world. Not all that exciting in the end."

"So tell me you're taking the job." Jack looked into Ed's eyes, who flashed that megawatt smile of his and looked away.

"Probably I will. I mean of course I will. Like I said, it's a gem. Everybody wants it."

"And you're the best."

"You said it. Not me."

"I've heard you say it."

Ed lifted his glass. "To the better engineer." Ed swirled the ice in his glass and raised the empty for the waitress to see. She came over, took both empties and orders for two more cocktails. "This job could be fun, Jack. Things we've only imagined but never done." He pointed at his friend. "Might even be able to get the new shoreline to store some carbon. A partnership between your people and mine. You and me."

"Now you're really teasing."

"You're so easy to tease."

"How long are you here?"

"Ten, twelve days. I've got to make sure they're really up for this before I close the deal."

In twenty minutes they ordered two more drinks, and another two. Ed warmed up. He turned his chair back toward Jack's. He touched Jack's shoulder, and twenty minutes later touched his knee. Jack was soaring. They ate like princes and carried each other out to the elevator and down to the street.

▼

It was a hot, damp evening and immediately Rex was in their face, waving a crinkled flyer. He was agitated, and his face was flushed bright red, white spittle gathered in the corners of his mouth. "They're moving us, " he cried. "Everybody goes!" He stank like the cheap whiskey he occasionally bought in pints.

Ed turned and pushed past the homeless man, ignoring him, but Rex came right back at them. "Oklahoma. Kansas," he said, waving the flyer. Ed raised a hand to push Rex back off but Jack stopped him.

"Hang on hang on," Jack said. "What is it, my friend?"

"Flood. Moving everybody out. All of us." He pushed the flyer into Jack's hand.

"Have you been there?" The guy went on, and positioned himself in front of Jack. There was thick white spit in the corners of his mouth.

"What's that?"

"Kansas. Oklahoma. You ever been there?"

"Not me."

"They say we have to go. Read it."

The crowd on the sidewalk carried Ed and Jack past the homeless man, who stepped off the curb into the street and called after Jack. "Everybody goes! That's what it says!"

Jack opened the flyer but Ed took it from him, balled it up and tossed it.

"Hey," Jack said. "I wanted to see that."

Ed laughed. "Space aliens. Second coming of the Buddha. Asteroids. End of days."

"You don't know that."

"Yes I do."

"He's a friend of mine."

Ed gave his old friend a look. "Nobody would agree with that statement. Least of all him." He put his arm around Jack and Rex was forgotten. "Now show me this casa of yours."

"Don't you have luggage somewhere?"

"It can wait. I'm looking to take clothes off, Jackie, not put more on."

▼

That night in his dream Jack could see the water rising. Jack was the water, and he could see the water. Then he was up in the building looking down at it all, pointing to the people down on the beach where he was also standing, but all he could see from down there on the sand was the small white face of a man in a faraway building, with a moving mouth, and no sound coming out. He woke in a sweat and slid his bare feet beneath the sheets to touch Ed's.

▼

There were plenty of flyers on the street the next morning and Rex was out, waving another one in the air, just as agitated as he was the day before. When he saw Jack, he took him by the shoulder.

"Take it in there with you," he said, nodding in the direction of Jack's building.

"To work?"

"See what they know."

"About what?"

"Oklahoma. People saying they're moving us."

"Moving who?"

"Everyone. All of us. Look buddy. I know what's going on. I'm the eyes and ears of this street. Eyes and ears. I know what I'm hearing and I know what I'm seeing. Kansas. Oklahoma. They're going to dump us. Get rid of us."

"No they're not," Jack said, and smiled. He'd already missed one cycle of traffic lights and when the light changed now, he stepped into the street. Rex must not have been able to read. "They're going to build walls," he called back, and waved his hand and the flyer in the direction of the sea. "All the way up and down the coast. I know the guy doing it."

Rex cupped his hands around his mouth and hollered. "Take it in there. Ask them. It's a flood, and everyone's got to go."

▼

Jack took the flyer with him down the sidewalk to his own building and took the elevator up to his office. He set the flyer on his desk without looking at it again. There were immediately three tasks, then four. There was a blowup in northern Europe where one of their tanks had leaked and asphyxiated seventy-four laborers. Three different media sources had contacted him wanting an explanation, an opinion, a word about risk, safety, liability, compensation. They wanted reassurance but they'd settle for a scoop. It wasn't until lunch when a colleague stepped into his office and picked up the flyer that Jack remembered it at all.

"Oh wow," the man said. "This is about the relocation plan isn't it? Is it starting?"

"It's one of those end of days things."

The man shook his head and studied it. "No it isn't. This is about the relocation."

"Relocation of who?" Jack got a weird feeling, a sense of the homeless man's eyes fixed on him, all the way down and out on the street.

His colleague laughed. "You look like a deer in the headlights. Not us. Them. Your buddy Rex, the street king."

"Who?"

"I've seen you buy him hotdogs. The homeless guy. The drunks and junkies and whoever else he hangs out with."

"Hangs out with?"

"They're rounding them up."

"Who is?"

The man shrugged. "They have to go, Jack. Another good storm and they'll be washing up on our doorsteps."

The hair went up on the back of Jack's neck. The last surge, seven years earlier, had filled the streets with churning mud, water, bodies, drowned pets, chemicals, shit, God knew what all—up to the second and third story of most buildings and houses. Anything smaller was smashed up, broken down, and washed away. It was horrifying, and the clean-up had been gruesome.

"There's a hundred thousand of them out there," his colleague said. "You've seen them. It's the decent thing to do if you ask me."

"So where are they going exactly?"

"Sanctuaries," the man said, and set the flyer back down. "You joining us for lunch?"

Sanctuaries. He'd heard of that somewhere. Jack shook his head. "I have to work," he said.

"Suit yourself. Want us to bring something back for you?"

"Nah, thanks."

He spent the next half hour looking around. The relocation had been approved in a referendum a few years back; the language was about humanitarianism and compassion, and now he remembered it. He'd no doubt voted for it. Sanctuaries, yes, complete with long-term housing, beds, lockers, and job training. Interstate agreements, federal compensations, it was all accounted for, legalized, locked in and ready to go. It had already begun in some cities.

▼

"But they aren't exactly voluntary, these sanctuaries," Jack said over a mid-afternoon cocktail with Ed. They were at a slick little martini bar halfway between their respective places of work. "Is that right? Participating cities are criminalizing failure to comply for anyone without an address?"

"Well," Ed said. "Jail's good business too."

Jack raised an eyebrow. "Seriously, Ed?"

"Think about it, Jack. This isn't hard. Picture this, ok? Between the shoreline and the city there's what, a ribbon of sand and green space about half a mile wide, right?"

Jack nodded.

"Ten years ago it was a mile wide. Ten more years and the water will be lapping up against the buildings, Jack. Think about that. And that's not even taking into account the storm surges. For now, that little band of green space is where most of these guys sleep, right?"

It was true. Years ago it'd been wide enough for playgrounds. A soccer field. Palapas. Cafes. A bike trail that snaked along the coast forty miles north and fifty miles south. A circuit of workout equipment: pull-up bars, knotted climbing ropes. All of that was gone with the last surge, and now the tide came two hundred meters further than where the lifeguard stations had once stood. Three dozen elaborate houses gone. It wasn't the community beach it had once been where people brought children and dogs and skateboards and bicycles. Now it belonged to Rex and his friends. You could go out at dawn and they would be curled up in coats and old sleeping bags. Some even had tents or make-shift lean-tos of scrap wood, metal or cardboard.

"So when the shoreline moves up to the city, permanently—we're talking one more storm, Jackie—where do they live?" He gestured out the window of the martini bar at the street. "Here?" Then he pointed a thumb behind him. "Or in-land, in the gated communities? In backyards? Or maybe with you? You going to invite them into your bed too?" Ed drained his glass and reached for a handful of peanuts.

"Shelters."

"You kidding me? Shelters? We're not talking two hundred people. There's over half a million in this city alone."

"Ok ok, I get it. But something about it seems wrong. Between the two of us, Ed. Is this really ok?"

The waitress put a new bowl of nuts between them. "Not if you ask me," she said. She looked at Jack. "I'll tell you something, it's a pretty fine line between pushing nuts and drinks and sleeping out on the beach. And I have a master's degree."

"Come on. With a pretty face like that I'll bet you could get whatever work you want." Ed winked at her. Jack's stomach turned and he looked away. It wasn't even clear what Ed's throwaway charm was supposed to be about. The waitress rolled her eyes a little but smiled and left them. Ed lowered his voice. "Master's degree," he said. "What do you think. Art History? Or no, Natural Resources Management."

Jack gave him a weak smile.

"Listen, Jack," Ed put a hand on Jack's shoulder, but Jack kept his eyes pointed outside at a planted tree. "There's a lot of shitty things going on in the world and this isn't one of them. It's really not. This is a huge population of destitute people getting a safe, clean place to live."

"You've been through the process before," he said, finally turning back to his friend. "Clearing everything out."

"I'm here as they're leaving. Not a coincidence. Right now, this is what progress looks like. Twenty, thirty years we might have a different solution. Right? We have to start with where we are." He leaned toward the blackboard behind the bar. "You sure you don't want to eat?"

Jack shook his head. "I got to get back."

Ed lifted his napkin from his lap and set it on the table. "I don't like that look on your face."

Jack finished his drink. "What look."

"Like I'm the bad guy. Jack. Come on. I'm not the guy who's behind all of this. I'm just trying to engineer some creative solutions. How is this not making the world a better place?"

Jack shrugged.

"While having some fun doing it, right?" Ed reached under the table and jostled Jack's knee, then moved his hand up Jack's thigh until Jack smiled and laughed.

"Ok ok," Jack said. "Maybe I'm over thinking this."

"Tell you what. Is there a decent place near the shore we could meet for dinner? There's some stuff I want to show you."

▼

When Jack took the elevator down at the end of the day, he thought about ducking out a sidedoor to avoid Rex, who he knew would be positioned out there, waiting for news. He pulled himself up, walked out the front, and sure enough Rex was right there, watching for him. Jack could smell the mustard and sweat and urine on the man.

"Well," Jack said, hands in his pockets. "You were right. It's the real deal."

"I knew it. I knew it." He pointed in the air to punctuate his enthusiasm. "It's a major flood isn't it? Like fucking Noah. And we all have to go."

"Well," Jack said. "Not exactly everyone."

Rex lowered his arm to his side. His grin slowly leveled into a thin-lipped, pinched line, his face wooden. Slowly, he started to nod. "You're shitting me," he said, very calmly. He looked Jack in the eyes. "You're fucking shitting me."

"Look," Jack raised his hands. "We have these models, you know? The water's going to come all the way up here, and it's not leaving. You've seen it happen once but this time it's for good. And there's going to be surges on top of that. This amount of water will kill you, but even if you survived, where would you go?"

"Where will you go?" Rex asked. But Jack didn't answer the question, and Rex did not expect him to. "I was in the middle of it, you know," Rex said, spitting a little and raising his voice. "I know what it feels like. You don't. None of you do." He stepped in close and stood rigid a few inches from Jack's face.

"Come on, man," Jack said. "I'm trying to talk to you here." Commuters on the sidewalk were watching them now.

"What are they going to do? Put us in cattle cars and ship us out, just like the fucking Nazis?"

"I don't think there'll be any trains like that," Jack said, but he imagined Rex and all the other homeless in handcuffs, ushered onto a line of buses. But that was ridiculous. Surely there were dozens of families who would be happy to be getting away from the water and would benefit from the assistance. "Think of the opportunity," Jack said.

"What are you, stupid?"

"Maybe it will be better in Oklahoma," Jack tried.

"Better than what?" Rex gestured behind him, to the strip of green and beyond that, the tattered makeshift dock and the sea. It was a gloriously sunny blue day. Rex pointed across the street to a stout guy with a beard in a filthy green and orange jacket. "See that guy?" The man waved at them. Jack nodded. "Known him fifteen years. Maybe twenty. You know anyone around here that long?"

"No."

"Course you don't," Rex snapped back. "You haven't even lived here that long, have you?"

Jack shook his head. "Going on seven years."

Rex ignored him. "When I'm down and out he helps me. When he needs coffee or beer I'm the one with the extra change. Five, six times this spring we did that for each other. We rode out the storm together last year. Him and me and a couple other guys. And a lady. Out here. Hanging on to each other.

"That's great."

Rex's eyes burned with anger. "It was not great."

Jack cut him off, already apologizing. "No, of course it wasn't. I'm sorry."

"And where were you, Mr. Office Building?" Rex looked at him, eyes inflamed.

"Don't you have any family?" Jack asked.

Rex looked away toward the water and for a minute didn't speak. "Got a boy somewhere," he said.

"Maybe you could find him? Maybe he'd have another idea about where you could go."

Rex laughed. His teeth were outlined with grey. "What, you mean you're not going to take me in?"

"I don't think that'd be a good idea."

"I bet you don't."

"Are you a vet, by chance?"

"Fuck you."

The old man waved a hand as if to dismiss all of it and turned away in his brown coat, leaving Jack alone on the sidewalk.

▼

Jack met Ed out by the water in a little pub—a local favorite designed out of artificial driftwood.

"Shrimp and suds," Ed said. He was already up at the bar licking grease and pepper off his fingers. "Let me order another round of everything and then we'll take a walk."

Jack raised a hand. "I'm not hungry."

"Have you eaten today?"

"I had two olives in my martini."

Ed shook his head and stood up. "You and your big worried brown eyes. Work up an appetite? Walk with me?"

"Sure."

Ed settled the bill and they walked out together on the pier.

▼

For all the change in the past ten years, it was still a pleasant beach, a place to walk when the light was changing and the birds were calling. Six months from now there would be a festival on the sand and grass along the parkway. City staffers, social workers and volunteers in red shirts would line up behind tables taking names, issuing instructions and bar-coded wristbands. There'd be free popcorn and lemonade, and there'd be thousands of home-less and unemployed men, women, and families getting their vitals taken in triage stations and being issued their health ID cards, many of them for the first time in their lives. They'd mostly be quiet. But there'd be a few jokers among them—tired old men and women with sores on their feet or in the mouths, who looked forward to the prospect of a comfortable bus ride and a little scenery, and promises of regular food and shelter. Blue and white buses would line up along the street, the rail shut down for the weekend. There'd be an air of holiday about it, but only because it'd be a big gathering outdoors by the water on a beautiful summer night; there'd be no music, and no one selling anything. There would be protests, though. Cameras and reporters. Small scraggling groups of men and women with poster board signs, accusing city planners of throwing away an entire population of human beings, supposedly for their own good. HOUSE THEM IN THE CITY, their signs would read, to which planners and residents would respond with words about the shore as a death trap, and the opportunities in the sanctuaries, which would eventually allow any able-bodied man or woman to return for a job, a permanent address, and a residence card. None of this would matter to Rex, who would be long gone, the sidewalk where he'd always stood as empty as a cavity, as quiet and unremarked as any square foot in the city.

▼

Jack allowed Ed to put his arm around him and turn him here and there, toward this bank and that. "Right here," Ed said to Jack, the little lights of the shrimp and beer hut blinking on behind him. "This will be the new dock, and there'll be twelve islands out there in an archipelago." He made a swoop with his arm. "Like a beautiful necklace across the water, with the most perfect mathematical precision to reduce the impact on the wall—which will be all the way up there."

"Quite an improvement on our dikes and ditches," Jack said.

"A few big islands, a few little ones. And people will build great little places out there Jack. Vertical gardens of oysters and scallops. Floating bars—ever seen one of those? Wait till you see one of those. There'll be boats you can take from one to the next. And the reefs will be engineered to move with the surges—come apart and re-assemble. We'll be able to control it all from a central computer." He made a gesture with his hand, then another. He started talking faster, smiling that Ed smile. He took Jack's hand and turned him 60 degrees. "Now I want you to pic-ture buttes over there. Fifteen miles long, a mile wide. We'll put shrubs and grass across the top. A blanket of wildflowers."

Ed went on. The city and shore will be de-signed to flood and even benefit from storm surges. You might have a beautifully tiled crater that's a plaza during dry times—a real work of art—and it stores runoff during storm surges. Or think, Jackie, he said, think of a tiered shoreline on one of the islands with buildings at the highest points, and lower levels prepared to take in water. He moved his hand to the small of Jack's back.

"I'll ask your advice on everything. Every part of it, ok? More your city than mine, right? We'll make it beautiful. Be my best work yet."

"It's sort of hard to follow," Jack said. "Will it do its job?"

"You never could keep up with my genius, could you?" Ed grinned.

No, Jack never could. He could picture it, though, and it did sound great, it really did. He could see it all glittering and populated with smiling beautiful people—all races, all genders, all ages. It would be a cutting edge, multi-use, health conscious and community-building public space. He took a deep breath, and interlaced his fingers in Ed's.

"Let's get some beer in you," Ed said. "You worry too much."

"It's not by choice."

"Sure it is." Ed gently bumped Jack's body with his own. "You'll warm up to all of this. You'll come around and see what a giant your old friend is. Do I know shorelines?"

Jack nodded and smiled, catching his friend's eyes.

"That's right. I do. And I'm going to fix a perfect little world out here for you and me."

THREE: HOLIDAY

The railless sped along at 700 kilometers an hour, smooth and silent as a silver wire pulled taut across the desert. On one side, the sea of misery and desperation of the refugee camps for which these three women—Gloria, Anya, and Clare— were trained. On the other, the domed city of New Harmony promising safety, pleasure, and relief. The women knew each other only casually in the camps, but their time off had coincided and they'd agreed to economize by making the trip together, anticipating a week-long vacation of drinks, hot baths, warm meals and ease. They'd all heard so much about life in a dome, but none of them had ever seen one.

Gloria—always frowning, arms crossed, eyebrows furrowed—was from the States. Rumor had it that she had gone to college and was a volunteer, but she was reserved enough that no one knew much about her for sure. She hadn't yet been in the camps a year, and was slow to find her place. Her roommate had said that Gloria cried herself to sleep every night her first month, and that after one day working in the nursery refused to ever go back in.

Anya found everything about Gloria offensive, down to her fastidiously knotted French braid. Anya had practically been raised in the camps, and nothing made her flinch, or sulk, or cry. For decades her parents had worked some of the camps' worst jobs: medical waste, sewage, laundry. If she'd had the resources to leave, if there were any refuge other than a different camp with all the same filth and sorrow, she would have been gone long ago. She was horrified—and a little fascinated—that anyone could have such a twisted sense of duty, or guilt, or self-righteousness, or hell, all three, to come half way around the world to work in the camps, for free.

Clare, the oldest of the three, had been displaced from her hometown outside of London when she was a teenager, and hadn't quite found another home until she landed in the camps. She was in charge of one of the main pediatric stations, and was among the first people to whom residents and newcomers brought their sick children. The night before they'd boarded the train for New Harmony, some twelve hundred new refugees had arrived. At dawn an impossibly skinny woman with hair like a matted cat's had placed a two-year old boy in Clare's arms. He weighed no more than a bird, his breath shallow, lips cracked, the darks of his eyes whitening, the whites of his eyes yellowing. She administered the 100 mg of morphine. She washed his face and head and chest, swaddled him in cheap, clean cloth, and within the hour had returned his body with condolences to his mother. Before she left, she'd seen a dozen children an hour between the ages of one and fourteen. This was her first vacation in seven years. She had seen to all the details.

Now on the train Anya shook out her capacious black hair and turned away from the other women, using her reflection in the train window to apply a bright red lipstick. "Here," she said turning back to face them. She opened a denim purse with bright snaps. "We should all have one of these. A toast." She took out three small vials. "I ordered them from Singapore. They're concentrated. All the drunk, none of the taste, half the expense. Perfect if you're like me and can't stand the taste of alcohol."

"Where'd you ever have alcohol?" Gloria asked.

"My great grandfather was an alcoholic."

Clare laughed. "Sounds like my family. London by way of Dublin by way of who knows where."

"He drank Daaru all day long. He always smelled like it." Anya closed her eyes and lifted her chin. "I remember it perfectly."

Clare held up her vial and read the label. Gloria turned hers over in her hand; it was small and cold. Anya unscrewed hers and held it out toward them. "To our holiday," she said, and flashed the beautiful smile for which she was known in the camps.

"To our holiday," Clare said, and tipped her vial back. Gloria did the same.

"I sort of wish," Gloria said, "that we could be going somewhere outside. I miss the trees and mountains. The sea. And there's so much of the world I haven't seen over here."

"Outside!" Anya laughed. "We practically live outside!"

"I'd like to see birds. There are no birds in the camps," Gloria turned from Anya to Clare. "Did you ever notice that?"

"Wrong again, Mother Teresa," Anya said. Gloria flinched and for a moment Clare thought the quiet American was going to cry. "We had plenty of birds when I was little. The kind that were after dead meat—and they didn't care what from, and it was near impossible to get rid of them. Is that what you want to see?"

"Oh no," Gloria said softly. "Of course not, Anya. I'm sorry."

"I'm with Anya," Clare said, and set her hand on Gloria's arm. "I have to admit, I'm looking forward to being—what?—enclosed." The thought gave her a visceral memory of being a child in her parents' house. There was a small brass lamp that used to sit beside her father's chair. She suddenly brought her hand back to her stomach. Christ. It hurt all the time.

"Just think," Anya said, and blinked her eyes closed to imagine it. "It'll be totally safe. We can just let it all hang out. No worries. None."

"Safe," Gloria murmured. "Did you guys hear what happened in the dome in Mexico last month?"

Anya put her hands over her ears. "Don't," she said. "I don't want to know."

"I think it's because they gave it international zone status," Gloria said.

"Would you shut up?" Anya snapped. "Please?"

"Sorry," Gloria barely whispered. She turned toward the window.

Clare took Gloria's hand and squeezed it. "Don't worry," she said. "We'll be fine. One week. Holiday. Right?"

Gloria leaned her head back and half shut her eyes. The contents of the vial were warming her blood, loosening her joints. Clare and Anya played two rounds of Virtual Smackdown and Lose-All-You-Have, their hands flapping like birds in Gloria's peripheral vision, their laughter shrill and sharp while the railless floated along at high speed. Outside, a blindingly bright world of white sky and hard dirt stretched as far as the eye could see. They passed a wrecked bridge over a dry riverbed, and the hollow shells of half a dozen abandoned cities.

"Those were trees," Clare said half an hour later, pointing out the window at the piles, bleached by the sun.

Anya thought there was probably something they could be used for—so much biomass—but she didn't feel like exerting the energy required for such thought or conversation. She interlaced her fingers behind her head and leaned back. You had to train yourself to disconnect when you had the chance. Compartmentalize. She knew some people who couldn't do it and it really messed them up. You had to know how to flip the switch. On, off. No in between. Yes, she'd left six hundred thousand refugees behind in misery, but she knew that if any of them could get away even for a day, they'd do the same. And what good did it do to be all twisted up with guilt like Gloria? Did she think it made her a better person?

▼

The train pulled into shadow and there they were, on the lower level of the dome. They slowed to a stop and for a moment, everything was quiet and still, the silhouettes of other passengers dark among the blue shadows. The orange interior lights blinked on and the passengers began standing and stretching, chattering as they found their luggage.

"All right ladies," Clare said. "Put your caps and vests away. We're starting with The Islands."

Gloria put her hand beneath Clare's elbow as she wobbled unsteadily. "Are you ok?"

"Something in those vials isn't agreeing with me," Clare said.

"Your stomach?"

Clare nodded.

"Do you need something for it?"

Clare waved her hand dismissively. "I need a champagne cocktail, that's what I need. With a slice of fresh fruit. Can you imagine?"

"Sounds divine," Anya said. "Let's wait till we have our toes in the sand and someone can bring them right to us. Have you guys ever been surfing? We should go surfing. Or skiing. Maybe we should go skiing."

▼

Advertisements for New Harmony had been accurate. Bring your hungry, your traumatized, your bored, your restless, your longing to forget. It was as green and vibrant a city as any of them could have imagined. There were lakes, gardens, schools, restaurants, shops, museums, and yes, an ocean complete with a necklace of islands and a pounding surf of beginner-, intermediate- and advanced-sized waves dissolving in a white lace of clean sea foam on golden sands.

In a small hotel overlooking the water, the women each had their own sleeping space and shared a common room. In each sleeping room, a bed, a vanity, and a small cooler. Anya opened hers and squealed. She stood in the doorway to the common room.

"Yuzu tea." She turned over the glass jar for them to see. "So expensive. I've only had this once in my life ever, but I loved it, *loved* it." She put her hand over her heart for effect, went back into her room to replace the jar and withdrew a small package of chocolate dipped rice wafers. "Oh my God," she called out. "My other favorite.

I haven't had these since I was a kid. How am I not going to buy and eat these?"

Clare and Gloria exchanged a smile. "I think it's always been the other way around," Clare said, "you eat them, and they charge you." She went to inspect her own cooler.

"Amazing," she called back toward the common room. She stood in the doorway and showed Gloria a small glass jar of pickled red grapes. "I can't remember ever telling anyone I liked these. Or even when I had them last."

"You must have mentioned it on-grid sometime," Gloria said. "That's how it works."

"What was in your cooler, Gloria?" Anya asked.

She shrugged. "I haven't looked."

"Gloria's beyond temptation," Anya said at Clare and rolled her eyes. She opened the chocolate rice wafers and stretched out across the rug on the floor. "God they really know just how to get you." She turned a wafer over in her fingers and popped it into her mouth. "I didn't know they still made them."

▼

The women swam in the sea and went shopping. More vials for Anya, a water storing jumpsuit for Clare. Gloria was interested in a sort of portable rain cloud she could've brought back with her to the camps, but she probably would've had to hide it there.

They settled on The Meat Lab for dinner—a famous restaurant high on their list of places to see. Their steaks were perfectly marbled without a trace of gristle or excess fat. Gloria tried to put aside the stories about how such perfect steaks were made, which wasn't very different from making babies.

"You are such a purist," Anya said.

"It's not that," Gloria said. "I just prefer the real thing."

Anya furrowed her brows and shook her head in mock confusion. "Is that not exactly what I said? Anyway, steak from cows? Seriously, Gloria?"

"Why do you say it like that?" Gloria shrank back. Anya's tone had been as harsh as if Gloria had said she killed and ate her own mother. Cow steak wasn't really that unusual, was it?

"This is totally real," Clare said. "Enjoy it." But she'd hardly touched her own food.

"I want a lobster tail tomorrow," Anya said. "Do they grow those here?"

"Rattlesnake, moose cheeks, bear belly, anything you want," Clare said. "They have everything."

"Excellent. Everything's what I came for."

"Do you think the people working here live here too?" Gloria asked, noticing that everyone employed there looked like Anya. "Like can they ever get out?"

"Well they couldn't live out there," Anya said, pointing at the wall behind her, toward the outdoors. "Anyway, why would they want to leave?"

"Why did they let us in through the basement though? Do you think there's people outside? Protestors? Like they say?"

"Like who says? Protestors? In the desert?" Anya laughed. "Fat chance."

"What would they be protesting about?" Clare said. "It's not labor protestors. Can you imagine an easier job?" The waiters and waitresses were clean and poised and full of quick wit and easy laughter, delivering plates of beautifully designed food to customers with money to pay for them. "I bet they live here. Can you imagine? Living here?"

"Sign me up," Anya said. "I bet there's a waiting list twenty years long."

Gloria made a face. "I'd like to see what's out there, on the other side of the dome." She stabbed at her steak.

"Course you would. Gloria the truthseeker." Anya gave her a flat stare as she sipped her drink. "She wants to look behind the curtain unlike the rest of us fools. You have to forgive us. We can't all be so wise and good."

Gloria's face flushed. "That's not it at all, Anya. I'd just like to see it. I feel a little out of place here in the dome."

"Oh you are," Anya said, chewing. "Everyone else here wants to have fun." Gloria looked at Clare to defend her, but Clare was pale, sweat beading on her upper lip.

"What's the matter Clare?" Gloria asked.

"I just." Clare put her hand on her belly and shook her head. "I don't feel quite right. I think it was something in those vials."

Anya leaned toward her. "Are you sick? Don't be sick!"

Clare shook her head. "I'm sorry. I don't know what's wrong with me. I'll be ok in a second. It comes and goes. It's already going." She smiled. Clare and Anya stared at her. "See? Better already."

Gloria put her hand on Clare's arm. "Do you need a doctor? Do you need to go back to the hotel?"

Anya pointed her fork at them, one at a time. "No! Stop it, both of you. Your belly hurts because you work too much and too hard. You don't need a doctor. You need fun. We're all going straight from dinner to Loverman, Inc.," she announced. "And nobody's going to be sick. Now. How does your belly like that, Clare?"

Clare laughed and took a sip of her drink. She took a deep breath and let it all out. "You're crazy."

"We are not coming all this way, staying for a week, and not trying it."

"I've heard," Clare said lowering her voice, "that they're not really machines." She wiped her upper lip with her napkin and set it back in her lap. "That the technology is a hoax and they're actually real men. But people don't like the idea of actual male prostitutes so they lie and call them robots."

"Dirty words," Anya said, running her hands up and down her body. "Makes me like it even better."

"Oh come on. They could never get away with that," Gloria said.

"But then of course I've also heard," Clare said, leaning in, "that the Lovermen are very, very high tech. They're continually taking your vitals as they touch you, and recalculating…"

Anya lifted a hand. "Don't tell me! I don't want to know."

"…to maximize your pleasure."

"Stop!" Anya said, laughing, and she put her hand over Clare's mouth. "I don't care how it works as long as it works. And we're all going. Clare's job was to organize the trip. My job is to make sure we have fun."

"What's my job?" Gloria asked.

"You'll bring us back to the stinking camps. Only person in the world I know who actually *wants* to be there. Isn't it true? You're a fucking volunteer?"

Gloria's face flushed. She set down her fork, set her napkin beside her dish, and lifted her glass. It didn't taste like real steak and they were all kidding themselves. And she didn't know what they made the so-called champagne out of but it wasn't grapes, or fruit of any kind, or anything real. All of this stuff was phony, terrible, and depressing beyond belief.

Anya withdrew a flyer from her denim purse. "Loverman, Inc. is a quarter mile away from here," she said.

Clare took the flyer. "Says it's off-grid."

"Of course," Anya nodded. "Very discreet." She glanced at Gloria. "For the Puritans."

Clare shrugged. "I'm game. Camp survival tip number one, every woman needs a dildo. How is this any different?"

"Just more advanced," Anya cheerfully agreed.

"I'm afraid to ask about survival tip number two," Gloria said.

"I know it's hard," Clare turned to her. "Being away when you're thinking about how much work there is to be done back there." She patted Gloria's hand.

"But I don't feel that way," Gloria said.

"Not at all?"

"No," she said flatly. "I just don't like the idea of fucking a robot."

Anya waggled her first two fingers at Gloria. "Don't restrict yourself to your own shallow experience, sweetheart."

Gloria swiped at Anya's hand, laughing.

▼

Clare pulled Gloria into Loverman, Inc. "Come on, see? There isn't even anyone else here." Anya withdrew a small flat device from the wall docks on which they were to answer questions about their preferences, sexual history, and health. Clare followed her lead.

Gloria fingered the gold charm on her necklace. "I don't think I can do this, you guys."

"At least see what they give you," Clare said. "You don't have to stay. You can leave if you want."

"It's a shame," Anya said to Clare behind her hand, in a voice perfectly loud and clear. "If there's one thing Gloria needs it's to get really well fucked. And it doesn't matter whether it's by a robot or a horse."

Clare grabbed Gloria by the arm to keep her from walking out. "Come on, just fill it out and see what they give you."

Gloria hesitated, and grudgingly sat down. She lifted one of the tablets out of its dock and began filling in the bubbles with her finger.

Anya finished the questionnaire first, and before she'd even returned the device to its dock, a tall man with brown eyes, stubble, and a dark ponytail opened the door.

"Anya?" he smiled.

Anya set her hand on Clare's wrist, and Clare grinned back at her. "See you girls in thirty minutes. Forty-five." She looked back at her Loverman. "Fifty?" They disappeared together through the door.

When Clare had gone, too, a small, slim man stepped out and took the seat next to Gloria.

"Scott," he said. "Columbus, Ohio." He extended his hand and they shook.

"From the States?"

"As sure as I breathe." He smiled.

"How about that." Gloria smiled, and nodded, and clasped her hands between her knees.

"Look, I get the sense from your answers that your friends brought you along for a ride you don't want to be on."

Gloria studied him. He was clean shaved and smelled like fresh mint. "You seem real," she said.

He laughed. "We try to transcend the binaries around here. Real, not real. You start talking 'real man' or 'real woman' and peoples' feelings are liable to get hurt."

"I—" she put her hand to her chest. "I'm sorry."

"We could just go get a drink, if you'd like, while you wait for your friends. There's a string of romantic tables for two around the corner. Or something more casual. Maybe you'd like a beer at the bar?"

"I don't know." She thought maybe his eyes didn't look normal. His hands and jaw and teeth were too even, too symmetrical. He gave her a perfect smile and put a hand on her forearm.

"Tell me. Where are you from? Let me guess. West coast? I know, I know. I can just tell." He shook his head and appraised her. "Lucky girl. Now what would bring you all the way out here?"

"You know what," Gloria stood up. "I think I'm just going to take a walk."

▼

She went past the stores and homefronts and ponds, beneath the trees and past the parks. There were three- and five-acre parcels of wildflowers, of woods, of rivers snaking and shimmering among low green hills that flattened out into another street of condominiums and shops. They'd left nothing out. It was like Noah's Ark. It was like a giant fish bowl. It was like all kinds of things, but it didn't seem particularly to be anything in and of itself. She hated it. How had she gotten talked into this? She had nothing in common with the people here. She sat on a boulder, thinking she had probably forty miles more to the edge of the dome, but then suddenly there it was right in front of her: the city ended at a doored wall just behind a row of globed street lamps. She looked around. No one was paying any attention to her, the street was lightly crowded with couples and families strolling and looking in windows and throwing coins into a fountain. She stepped behind the lampposts, opened the door and stepped right through.

It felt better on the other side of the door, like she could breathe. God, it was awful inside that thing. She followed a concrete staircase down one, two, three, four, ten flights of stairs in all, whereupon she pushed against another door beneath an exit sign and stepped all the way outside.

It was like stepping into an oven, searing daylight all around, and she realized she didn't know what time it was. She'd thought it was nearly evening—hadn't people in there been eating dinner?—but out here, it felt like noon. What latitude were they on? Directly before her was a broken slab of asphalt and a pile of metal and wire trash. Far off in the distance, the white chalky line of what she thought might be mountains or hills. A few twists of dead mesquite, or something like that, and leafless trees like upright bits of frayed twine. Other than the wind scouring her shins and arms with grit, there wasn't a sound. Blood rang in her ears. She stepped all the way out and swung open her arms and tilted her face back into the sunlight. It felt holy out here, it really did. The place was exactly itself. The door swung shut behind her, locked. Good, she thought. Never going back in there again.

▼

"I'll tell you," Anya said. She was leaning back in a waiting room that was inaccessible from the front entrance, her legs stretched out, her hands interlaced behind her head, long black hair pouring out behind her and over her shoulders. "Somebody take a picture. I could be an advertisement for this place."

"You look thoroughly ravished," Clare said, laughing, and slumped down next to her. Her hairline was damp. "What do you say? Human or machine?"

"There's no way that thing was human. I can't even walk."

"I know. When's our next vacation?"

"Vacation, hell. I was just figuring if we got the railless right at the end of the work day, we could get here in time for one or two turns and be back in the clinic in time to feed the miserables. Speaking of which, where's our sister of mercy?"

"Maybe she asked for seconds."

"Oh hell," Anya said. "She's not in there. I bet she went to find a church or something."

They laughed. "I'll let her know that we're finished," Clare said, retrieving her PCD. "Way finished." They stepped outside and Clare pointed to some high bar stools on a balcony strung with tiny globes of neon blue light, pouring out laughter and music from old movies. "Want a martini?"

"Yes. With big green olives from a little glass jar. Shaken not stirred." She put on her most sophisticated face as Clare laughed.

▼

Outside in the sand and sunlight Gloria kept within arm's length of the dome and began walking toward what she figured was south, since the sun was up to her right. She'd expected a whole service industry of people out here, and maybe refugees and protestors and soldiers and police. Shouldn't there have been people objecting to something? To what, she couldn't have said—but if she'd been cast out of the livable world and stuck here she'd be protesting *something*, that was for sure. She walked twenty minutes and somehow lost sight of the dome. How was that possible? Twenty-five minutes, then thirty. Her throat was parched. She could feel her brains baking behind her forehead.

▼

"Forget her," Anya told Clare and reapplied her lipstick as they walked. "She's probably back in the hotel. Or else halfway back to the camps by now."

The two of them were walking along the sidewalk between the rows of shops and restaurants, green masses of leaves shushing and shifting on the tree boughs overhead. Clare pointed up, one hand tucked into her waistband over her aching stomach. "Look. The Big Dipper." There it was above them, with a depth that made it seem light years away. "How do you think they made that?"

"How did they make it? How do you even know what it is? Big Dipper?"

Clare lowered her arm and drew her sleeve across her forehead. She was really sweating. "I grew up far away in a place that was cold and dark at night. I spent a lot of time looking at the stars." She pulled out her PCD again. Still no word from Gloria. She was getting worried.

"And then what," Anya said, and there was a new, hard edge in her voice. "You decided to leave your little heaven and save the rest of us?"

"What?" Clare looked up at Anya. "No. It wasn't quite like that." She slowed her pace and winced.

"What was it then?"

"I need to find a bathroom," Clare said as they passed the turn that led to the ski mountain. "I think I'm going to be sick."

▼

"Sweet Jesus," Gloria said. Her hands and fingers felt swollen at her sides. Grit in her teeth. "Am I going to die out here?"

"Nope," a man said, coming up out of no-where on a tiny, ultra quiet, rubber-wheeled car. There was a river of dust floating just above his tracks in the distance. "But you are in trouble."

She whirled around and instinctively took a step back. "Oh my God, you scared me."

"Anyone who walks straight out of that," he said, nodding at the dome, "and into this? Nothing but trouble."

"I got locked out."

"Get in," he said. "I'll take you around and back down to the ground level."

She climbed in, grateful. "I just had to get out of there, but I got so disoriented."

He had on a smooth, wide hat and was fully clothed from chin to boots. A mesh screen draped down from his hat before his face. She couldn't tell what race he was, or even what color his eyes were. "It's what I'm out here for. We get about a dozen a day."

"A dozen what?"

"Nostalgics."

She shook her head. "I work in an aid camp hundreds of miles from here."

"Miles? You from America?"

"Yeah."

"And you like that camp work?"

"I hate it."

"Why are you doing it?"

"I thought I was supposed to."

"You come here with friends from the camp?"

"Coworkers, yes."

"Don't like the dome either?"

"No."

He laughed, then shook his head and reached into the back and pulled out a heat-deflecting blanket. "Here. Put this around your head and shoulders."

"I don't know how you can stand it out here all day," she said, draping it over her head.

"I'm a robot."

"You are?"

"No. I'm one of those aliens that work outdoors in the sun and wind. They had to go to another planet to find me," he joked.

"I work outdoors in the camps."

"Not like this you don't."

"No, but it's not much better, either." There wasn't quite this level of exposure, but there was the sun and heat and tedium. The never-ending tasks of feeding, treating, vaccinating, cleaning, shoveling waste, wringing out laundry. The mouths, the hands, the eyes; the tears, and creams, and wipes; on and on, round and round... "I could have died out here."

"Not a chance. We catch everyone," he said, and pointed up, where she could see an angled mirror.

"Mirrors?"

"Cameras. And lasers fifty miles out," he said, nodding west, "if you made it that far. Everything is patrolled."

"Why all the security?

"We like to keep track of everyone."

"Who is we?"

"By we, I mean they."

▼

Clare vomited down the front of her shirt before she made it into a stall, then threw up again in the sink. "I'm so sorry," she said, her eyes filled with tears.

"It's ok," Anya said. "Come on. Come on." Clare lifted her arms and Anya pulled the soiled shirt off over her friend's head. She set it in the sink and turned the tap on. "More?"

Clare nodded, put her hand over her mouth and ducked into a stall. Anya rinsed and soaped and wrung out the shirt. When she turned off the water she could hear Clare coughing and spitting. She hung the shirt on a hook behind the door and stooped down in the front stall.

"You're really sick."

"I'm sorry."

"Can I come in there with you?"

"If you want." Clare reached back and un-locked the door and Anya came in and kneeled beside her. "I thought if I could just get away for a while. Just relax a little, you know?" Anya could see snot hanging from Clare's nose. She tore off some toilet paper and gave it to her.

Clare nodded into the tissue and wiped her nose and mouth.

"I'm just going to rub your back," Anya said. "Ok?" Clare's bra straps were filthy.

"Ok."

"Was it something you ate?"

"It hurts all the time."

"Before today?"

Clare nodded. "All the time."

"How long?"

"Couple months."

Anya shook her head. "Have you been to a camp doctor?"

"It didn't seem serious."

"Well it does now. Right?"

Clare nodded.

"Why didn't you say something?"

"To who?"

"Ah, Clare. There are people who can help. You need to take breaks. You need to take care of yourself."

Clare nodded and wiped her mouth with the back of her hand.

"We need to go back," Anya said. "Ok? I think you need a nurse or a doctor. One who knows what kind of stress you're under, what you've been exposed to."

Clare breathed, her back rising and falling beneath Anya's hand.

"It's pretty bad in your station," Anya said. "Isn't it?"

Clare nodded. "It's bad," she said. It came out in a hoarse whisper. Then she let go, and really cried. For several minutes the women held each other on the bathroom floor, Clare's hand trembling on the toilet seat.

"You need to be in your own bed," Anya said quietly. "You need to talk to someone who knows what to say."

Clare wretched and coughed again.

Anya made smooth, slow circles on her friend's back. "We can come back here some other time," she said. "No big deal. Right? No big deal."

Her eyes and face red, Clare turned back to look at Anya. "I'm so sorry."

Anya rocked back on her haunches and sat on the floor, her own eyes filling with tears. She wanted to stay here in the dome so much. Just a week. Just one stupid blessed little week.

▼

Gloria and the man rode on in silence until he pulled the small car underground into shadow.

For a moment she was blind. It was instantly thirty degrees cooler.

"Where am I taking you?" the man asked. "Do you remember the name of your hotel? Which of the main quads you entered?"

"How do I get out of here?" Gloria asked. "Isn't there like a central station under here somewhere?"

"What, you're not going back in?"

She made a face. It was the same one she'd made over the steak, and when she'd laid eyes on the artificial surf, and when she'd walked out of Loverman, Inc. It might have been the same face she made when she first entered one of the "nurseries" at the camps. "No."

"Back to the camps?"

She shook her head. "I'm going home."

"To the States? From here?"

She nodded.

He whistled behind his mesh screen. "It's not easy to get home, sweetheart. It'll cost you."

"That's not a problem."

The man turned away from Gloria and pointed his gaze down the railless line. He felt his face harden. Just like that, she'd get on a couple of trains, a couple of planes, and be a world away. Probably in a private house. A family. A garden. He closed his eyes a moment then turned back to her.

"You don't have any luggage."

She waved her hand dismissively. "It's all camp junk. They can have it."

He pulled the screen back off his face and looked at her with naked eyes. His skin was dark and lined with sun and age, his eyes very blue. He looked right at her. "Won't your friends be looking for you?"

She climbed out of the cart and brushed herself off. For a moment the dust sparkled in a plane of harsh sunlight, then disappeared in the cool shadows.

"They're not my friends."

FOUR: SHANGHAI

He stumbled out of downward dog, turned around, and saw her. He was used to smiles, especially at yoga, but this was no ordinary smile. It was so magnetic he felt himself sliding across the floor like an iron filing.

On the crowded elevator down to the street, arms and shoulders pressed against another dozen bodies, she smiled again. He reached past the man beside him and she reached up to take his hand.

"Hi," he said. Her skin was smooth.

"Hi."

There was some chatter in the back corner of the elevator but no one else spoke.

"I haven't seen you before," he said.

"Just visiting."

"Work or pleasure?"

"Work."

Outside, the street was loud and the sunlight muted, but the wind was bracing. She pushed her dark hair across her face.

"What are you doing right now?" he asked.

"Hungry."

"Walk to the market?"

They fell into step beside each other and he was aware that his smile was a little too big. Meeting someone in person like this—a spontaneous magnetism like that. Nobody got to have that—nobody. It was like Cary Grant and Deborah Kerr. Or like where the beautiful woman decides to get off the train in Vienna to follow the handsome American man. Or no, no—he glanced at the woman walking gracefully beside him with her chin up and dark hair poured down her back—she's a princess from an unspecified country and he's found her out, and he's going to follow her to the ends of the earth. Claim her for himself. He gave her his elbow. She looked down at it and laughed at him, but hooked her slender arm through his.

They walked the streets of his city, the air thick with humidity and energy, and raised their voices over the din. He asked her about work, and she told him a little about each of the cities she'd recently visited, and ten others besides.

"Do you have a favorite place?" he asked.

"No."

"A home?"

She laughed at that.

"What do you do when you're not working or orbiting the planet?"

She shrugged. "Feelies, smellies, anything I can get lost in."

They stopped behind a line of bodies waiting to enter the guarded square where the market stands were assembled.

"Do you need to get lost?"

"I need to have fun."

"VR fun?" he said, with a half smile.

She glanced at him and raised an eyebrow.

"Seriously," he said. "Real or virtual?"

"Obviously that depends."

"In general, though?"

She shrugged. "In general, there's no difference."

"You do not believe that."

"Yes I do. And so do you."

He furrowed his brow at her in mock frustration as they stepped up closer to the gate. "Wait till I show you the tomatoes," he said. "You'll change your tune."

She rolled her eyes but gave him a beautiful smile. "I've seen tomatoes."

"Some of these were grown in gardens."

"No," she said. "That I do not believe. But I do believe they'd price them as if it were true, and that silly romantic men would pay for them."

"Are you calling me a silly romantic man?"

"Have you bought those tomatoes?"

The market was crowded with musicians and craft stands and tables stacked with shining fruits and vegetables. He looked at her.

"See?" he said.

There were the tomatoes—big, red, hydrated tomatoes. He waved one beneath her nose. "When's the last time you smelled that? Ever?"

She looked at him. "It doesn't matter what form the tomato comes in."

"What? Didn't you ever smoke a real cigarette?"

She shook her head and smiled at him as if he were either a child or an old, old man.

There were even avocados. When he was a kid, he told her, he lived near the avocado groves and imagined himself a farmer. The drought had killed them off.

"Oh my God," he said suddenly. Apples. Big, green, sour, juicy ones. When he saw them, he stamped his feet in a percussive shuffle. She watched him with amusement—he was kind of cute—but she was indifferent to the apples.

"What's the matter with you?" He asked, setting one of the apples back on the stand. "What's your favorite fruit?"

"Cherymoya, Rambutan, Miracle fruit, passion fruit. Sweet. Anything sweet."

"Ah, we're back to fun, aren't we?"

They wandered through the market and he bought her a mango. She turned it over in her hand.

"Where do you think they grew this one?" she asked, half-teasing. "An island two thousand miles away and flown in just for me?"

"Somewhere here in the city," he nodded in the distance toward the buildings. "But we could pretend it was a mango grove."

She gave him a funny look. "Why would we do that?"

"Just tell me you'll enjoy it."

"Of course I will." She began peeling the fruit with her manicured nail, expertly pulling the soft stringy pulp and handing him a bite with sticky fingers. Juice ran down the inside of her arm.

"Most people need a knife for that," he said.

She took a bite of the mango. Her teeth were even and clean, and the fruit made her lips and chin glisten.

They circled among the people and passed a woman with two children. "Want one?" he pointed. "I'll buy you one," he joked.

She laughed. "God, no!"

"Not ever?"

She shrugged.

"If you did, how would you do it?" He touched his belly, looked at hers.

She waved a hand. "I guess if I wanted one, I'd just get one."

"Do you want a family?"

She shrugged. "It's a little tribal. And really not good for women."

"I have a couple friends with families. They seem ok."

"Religious?"

He considered. "I guess you could make that argument. But not in a God kind of way."

He told her about visiting them, these friends: a man, a woman, their son and their daughter. How in the morning they'd put the kids at the table, and the table was all crusted with finger paint and pancake batter. And they'd sing before they ate—a different song to break the fast every morning.

"Oh my God," she said, laughing. She grabbed his arm. "You are so full of bullshit. That was in a book—what was it called?"

"Did you read it?" he asked. "Did you love it? I loved it."

She laughed and shook her head. "Why would you just lie like that?"

"I thought it was nice. I thought it sounded nice. They stay married like that for I don't know. Forty years. And there was the thing about the son—and how the daughter had to save him."

They came back out on the street where they'd entered. It was twice as crowded and the sun was a bleary white eye behind the low hanging sky.

"Don't you get tired of being in the same city all the time?" She asked.

"The same city?" He pointed to the windows as they passed. "This is where all the art and fashion come from." They both looked into the shop window they were passing. "And all the newest wonders."

"No," she said. "I've seen those cities. This is not one of them."

"So critical."

"Realistic."

"Lovely."

She turned her face up at him and smiled. He touched her cheek and drew his finger along her jawline.

She asked about the creatives. He admitted they were thin on the ground. "The city is more of an aggregator than a generator." He thought she seemed disappointed to hear it.

"That was one thing I thought this place maybe had going for it."

"Well," he said. "Maybe it's just in a rut. But I'll tell you. One of the things I really love about this place," he said, "is sleeping in a city that's always awake. It's like, did your parents ever have a party, and the sound of their laughter downstairs, and it's late, and you're up in bed?" He caught her eye on the last few words.

She felt her face heat up. "No," she said. "My parents never had parties."

"It's like you're home safe," he said.

She imagined what it would be like with him.

"And someone's looking over you," he said.

She tied her hair back with a sudden movement. "I'm not sure I get it," she said.

"Next time you're in town," he said, "I'll show you." He nudged her with his elbow, and her body gave a little. "You can stay with me."

They walked close, their upper arms brushing each other, and separating, then touching again. In front of the station he leaned in, and she extended her cheek, and left for the jetport.

▼

When his VR caught her she was on a train. Her face was flushed, her hands were clammy, and her chest heaved.

"Where are you?" he asked.

"I have a stomachache," she said.

"You're the only satellite I know who suffers from motion sickness."

She moaned.

"Want me to rub it?"

She glanced over at the guy chewing on the end of an electronic stogie. Not a flicker crossed his face.

"That would be nice," she said, and took his hand and placed it where her stomach hurt. He started with the big rumbling around her abdomen, then his strokes became longer as she began to relax. Suddenly she sat up and looked grateful. She put her hand on the back of his. They both smiled.

"Shanghai," she finally said.

"How long will you be there?"

"A few days," she said.

"What will you do?"

"Meetings," she said.

"I mean for fun?"

"There's a restaurant I like. We're going to-night."

"The one you told me about, where they dress you up like the Ming Dynasty?"

"That's the one."

"I want to see you with a tower of black hair and pearls."

"It's really heavy that way."

"Silk robes. Waving big paper fans."

"You'd just make fun of me."

"I would not."

"Oh," she lowered her voice to imitate his tone the night before, "'What's the Empress of Fun doing tonight after work?'"

"I was just teasing. I like to have fun too."

"Don't tease me."

"Ok. I'm sorry. Tell me what you'll eat at your restaurant."

She put her hand to her mouth to stifle a laugh. "Fattened ground squirrel."

"No."

She nodded. "They put on a furry tail and chop it off in front of you."

"Do you get to keep the tail?"

"I think it would rot or fall apart. Anyway it's synthetic."

"What else?"

"I love the steamed cow's milk."

"Can you get it sweetened?"

"And flavored."

"I had real cow's milk once when I was a boy."

"Where?"

"California. My uncle lived near some of the old farms."

"That is really incredible."

"My claim to fame."

"What was it like?"

"Kind of slimy, I think."

"That must be one of the features they re-move."

"Will you be with friends tonight?"

She smiled. "Work friends."

"Where are you staying? The French Con-cession?"

"French Concession!" She laughed. "You might as well call it the French depression. Or Eurasian deception, conception, decapitation. Where do you get this stuff?"

"Sorry sorry," he laughed. "Everything I know about Shanghai I get from old movies."

She rolled her eyes, then squeezed them shut. "Getting a headache."

He started rubbing her belly again, and she drifted off. The man with the stogie caught a quick look, smoothed his slicked back hair, felt for something in his pocket, and returned to the screen floating before him.

▼

Later they met on a beach crowded with machines—
the kind that leap up, dive down, do back flips, or
placidly float, all when you least expect it. Mainly
what they're good at is getting shrieks out of people
who have never seen the ocean—or anything like
an ocean. The sky was a perfect enamel blue.

"Who made that sky?" he asked, looking up,
then shaking his head in wonder and appreciation.

She shrugged. "Who cares. It's perfect, right?"

"Oh, look, rats," he said, pointing across the
beach.

"Who put those here?" She made a face.

"They're huge!" He loved it when there was
a flaw, or a mistake, or something surprising that
broke up the set a little bit.

"Let's try over there." She pointed. "It's rela-
tively clear."

"How about we just leave all this behind and
colonize a new planet."

"Ha," she said. "That I could get behind."

"What'll we call it?"

"Yours and Mine," she said.

"How about just Mine? But I'll let you live there with me. After all, you'll have work to do on planet Earth. Unless you give it up to settle down with me."

"I'm not giving it up."

"You don't want to be my love at first sight? My beloved peach? My lady all ripe and big and fat with our babies?" He ran his hands over her breasts, he kissed her neck.

Her arms hung at her sides. "You don't want any of that," she said. "You just like the story."

"You're wrong about me. And you're wrong about this. I'll get you." He wrapped his arms around her waist and pulled her in close. "I've already got you."

They spent a lot of time in bed—sleeping, making love, watching movies, feelies, and smellies. They were staying at Zorba, the Greek-themed hotel. They rode bicycles along the water, imagining they were simple Greek fishers from the last century—or Paris bohemians from whenever. They ate when they felt like it and the food was whatever they wanted it to be.

One afternoon they hiked up to the hot springs in the mountains. It was designed up to make you feel like you were at a religious shrine. Little Buddhas and lotus leaves here and there. Around them the manufactured clicks, whirs and whistles made to sound like insects and birds. The density of wide green leaves waved like a hundred hands.

"No," she said, when he took her around the waist. "Not here."

"There's no point in your being old fashioned. Everyone can see everything we do, you know."

"Just stop."

"Everywhere we do it."

She pushed him away, but smiled.

"All the time," he said. "Like last week when we were on VR in Las Casas?"

"Stop!"

"New Delhi. Remember that? Everybody saw that."

"They did not."

"I bet we were somebody's big entertainment extravaganza. Romantic man meets and seduces coldhearted woman."

She punched him lightly, but she knew it was true. The guys running the software could take her top off with the flick of a switch. But then it occurred to her—maybe there is no guy running the software. Maybe it's software all the way down.

"Another reason not to worry about modesty is there's no privacy anyway." With that, he took her hand, and they finished their walk circumscribing the sacred grounds before reaching the hot springs.

"How long do we have?" she asked.

"Ninety minutes."

She stuck her toe in and frowned. "This isn't hot."

"Try it over here."

"Scorching."

"I'll change the settings."

She looked around. "Are we just supposed to sit on the stone?"

"Does my lady want a divan?" he joked. "It's a natural hot springs. The water comes up out of the rock."

"Ugh. Whose idea was this?"

He offered her a massage. She shrugged him off, then draped herself over a rock, feet in the water, making herself as long as possible. He couldn't take his eyes off of her. Later, he would think of this moment again. Her figure in the water. Her simple complaints. Her clarity. Her efficiency and speed. It was like she was from some other planet. She made the women in his own city seem like warm, sloppy animals. There were hundreds of available women, just as he was one of hundreds of available men—all of them in an elaborate line dance changing partners. That's how it was. But he didn't want that anymore. He wanted the real thing. He wanted her.

▼

On their last night they had dinner at the Hotel Nacional de Cuba. He pulled her chair up right next to his.

She laughed. "Can you get any closer?"

"We could adjust our settings and I could get you right up here on my lap."

She slapped at him playfully.

The waiters wore white gloves and shimmied silently through the room to the electronic strains of the son montuno, mambo, guaracha, guajira, cha cha cha, afro, canción, guaguancó, and bolero. The walls were decorated with murals depicting parrots, peppers, writers, revolutionaries, gangsters, and politicians. They drew close together against the sound, picked at their food, and spoke softly but quickly.

"I'll see them in two days," she said. "All those sick people just over the border." She saw something like fear in his face. "I won't be near them," she reassured him. "I'll just be able to see them from a distance, where they live."

"It's amazing, the people who really go into those places to work."

"The doctors and nurses? They all get sick too. Or they already are."

"Promise me you won't get sick."

"I won't get sick," she said.

"You know," he said, "in the future they'll say we didn't do enough. Especially people like me. I do nothing."

"There's nothing else we can do. Make people comfortable if we can. That's it."

"Don't you worry that we're in the midst of something bigger that we can't see?"

"You see what's in front of you. You do what needs doing."

"A lot of people wouldn't do it."

She shrugged. "It's work. It needs to be done, and I like doing it," she said and squeezed his hand.

"Let's figure out when I see you again."

"You see me now."

▼

They had gotten so good at controlling the weather that nobody took much notice anymore, except when the systems crashed. Usually it was just a matter of hours or days until they were up again, but this time it was different. The storms had been going on for weeks and it was getting hard to go out. Most of the remaining trees were horizontal. Water was bubbling out of the sewers along with God knows what. He had heard that a hospital had been forced to evacuate its patients. He felt safe, but was starting to get cabin fever.

He looked out the window at the city and took a long breath. He walked to the kitchen and pulled a beer out of the fridge. No matter how bad things got, there was always food, drink, fun and games, even if the beer was sometimes lukewarm. That was the social contract. He punched some numbers into a device and she popped up on the screen, looking lovely as ever in a white T-shirt.

"What's wrong?"

"Bored and lonely," he said. "But mostly wanting you. My one and only. My sweetest girl. What are you doing?" he asked.

"Working on my deck. I've got a meeting in an hour about a food manufacturing plant."

"What's the deal?"

"If we can pull it off we could reduce caloric deficiency in the region by one or two percent."

"Not much," he said.

"That makes a big difference, especially if you're a kid or a nursing mother."

"I can't believe mothers still nurse their young."

"It's pretty amazing."

They chatted this way for a while. She smiled a lot and commented on the weather outside the window behind him.

"It sounds like we've got enough time for a little sex?"

She grinned. "OK. Just let me slip into my VRXXX Premium."

She was really something. Everyone else just went at it, but she had invested in top of the line VR encryption technology. It was both endearing and baffling.

"It's just for us," she'd said. "Nobody else gets to see."

"You're from another age," he teased. But he didn't know if it was an age that had already passed or one that had yet to emerge. She had a way of making him feel behind the curve no matter what she did.

He caught a quick glimpse of her on the screen, and then it went blank. At first he thought that she was gearing up, but nothing was happening for too long. He felt his desire starting to go into remission.

Suddenly the screen caught fire and she appeared in an Italian Contessa negligee. He reached out, kissing her slowly and caressing her neck. She nibbled on his ear. Just as he began tugging the string on her negligee, she started breaking up. Suddenly his hands were in a vacuum and her touch became a series of random jabs. Then she froze completely. When she came back to life her voice was distorted.

"Maybe we should try another channel," he said.

"It won't be premium encrypted so we'll have to take it easy."

"All right."

He punched in another set of numbers and she appeared again in her white T-shirt. They embraced. He could feel her arms around his shoulders, slowly sliding down his body. He kissed her, running his fingers through his hair. He knew what he wanted to do but he was afraid she would object. He could feel her touch but it felt far away and tentative, then he lunged for her and the screen went blank.

"It's not working," he said to himself.

After ten minutes of fiddling a clear image finally popped up on the screen.

"I'm sorry," he said. "This is just not going to work."

"Oh. Ok. Well I have my meeting soon."

"I miss you," he said.

"I'm right here," she said. "I'm all yours."

"I know."

He said all the right things, the screen went blank, and he sank on to the couch.

He thought about going outside but he gave up on that idea after a quick look. He tried reading but he was restless. He put on some music, and was soon bathed in Miles Davis's "Kind of Blue." He woke up with a start, unaware he'd fallen asleep. He had no idea what time it was but he was hungry, and the lukewarm beer was now sticky and sweet. He poured the beer down the drain and found a package of ramen. He turned on the water, but only a trickle came out. He put it on the stove to boil. He and the stove stared at each other. There wasn't much energy in either.

He sighed, and looked around the apartment. It had been a long time since he had talked to anyone in his building. His life was elsewhere.

He wondered if there might be a new pub or inn or restaurant or anything within walking distance, then realized he didn't care. And anyway they might not be open. He didn't know which systems were up.

There was a woman who lived on his floor. He had seen her the other day by the compactor. Funny he hadn't noticed her before. His stomach growled and he turned off the stove. He put on some slippers, picked up the ramen, and shuffled down the hall. He rang the girl's buzzer and waited. Finally she came to the door and peered at him through the peephole. She looked surprised. He held up the ramen.

"Do you have any way of heating these noo-dles?"

She opened the door to the chain's length and carefully looked him up and down from head to toes. She raised an eyebrow. "Is that really what you want?"

He nodded.

She smiled and undid the chain.

FIVE: ZOO

It was so old-fashioned: a blind date arranged by a common acquaintance. She was two years a widow; he, a forty-five-year-old bachelor. She hadn't been game for meeting up until she found out what he did for a living. Even their meeting place was a throwback: they were tucked into the back of a small, chic restaurant in the heart of the city. They were at a table for two that was so small their knees kept touching beneath it. He turned his chair sideways slightly, took a sip of his drink and shook his head as he swallowed. "God," he said, and laughed. "You sound just like the school kids."

"No," she said and smiled, and crossed her legs behind the table. "Tell me. I want to know what it's like. Do you pet her? Just like, hang out with her?"

"I usually wait until the second date to talk about the zoo." He winked, and grinned at his own joke; she just stared at him. She was almost beautiful, he thought—her eyes just slightly too close set, her nose just a little long. He liked her perfume, though, it reminded him of a citrus blossom—something he hadn't smelled in decades.

"Honest," he said. "It's not as exciting as you might think."

"My father saw a bear once."

The man shook his head and smiled. "No he didn't."

She smiled back politely, but with her eyebrows furrowed and head tipped slightly as though she didn't understand. "You're saying he lied?"

"I'm saying there's no way he saw a bear. He must have mistook something else. I mean, around here? Recently?"

"This was ten or fifteen years ago. About two hours east."

He shook his head again. "I don't think so."

She nodded, then lifted her drink and gazed across the restaurant. As the kitchen door swung open behind her, the light within momentarily outlined her profile—chin lifted, nose long and straight. He cleared his throat, opened his mouth to speak, and closed it again. This had been a bad idea from the start.

She set down her glass and crossed her arms, studying the wall across the room, behind his shoulder. The tables in the room were all full, the lights dim and rosy. It smelled like an expensive potpourri of garlic, white wine, eucalyptus and roasting meat. There was a small trio of piano, muted sax and bass coming from the far side of the room. The pleasant chime of silver and glassware.

"Anita," he finally said. "That's the tiger's name." She glanced at him and he raised his eyebrows: did she want to hear about this? She turned her shoulders and faced him.

"I'd be so interested to hear about her," she said, and picked up her drink again, and leaned a little forward. "Is she really the last, last tiger?"

"Actually no," he said. "There's a Bengal in London, a male, but he's eighteen—even older."

He told her what it was like—what the kids and zoo visitors always asked, and what he told them. They always wanted to know if he and the tiger were best friends. Like, did he ever get in the pool and swim with the tiger? He always answered the same way. First of all, no. He and the tiger were not best friends. Second, no, they did not swim together. That would be terrible for the tiger and probably not very good for him, either. Sure the tiger was cute, or beautiful, however you wanted to say it, but she could hurt a person, even unintentionally. Anita weighed in at over three hundred pounds. She needed special, expensive medical care by highly trained veterinarians, twenty to forty pounds of fresh meat per meal, and a lot of space. And she was old.

"So you see," he said, shaking his glass to shift the ice around in his drink. "Not that interesting. I'm her caretaker, not her friend. Ideally, a tiger would want a tiger for a friend, just like a girl wants another girl for a friend."

"But there *are* no other tigers for this tiger."

"Nope," he said.

"God." She shook her head. "It's such a tragedy."

He shrugged. "Tragedy? I don't know about that. This is what we do. Kids get to run across the so-called savannah and boreal forest with a dozen virtual look-alike big cats, the zoo saves money on food and maintenance and vet bills, the city saves millions of gallons of water per year. Animal rights activists are finally happy—"

"I'm sorry, because why?"

"No animals in cages. Very few anyway."

"Ah, ok. Wow." She made an exaggerated frown, clearly in disagreement with this logic.

"Anita's a very spoiled, very comfortable cat, and everybody's happy. Happy happy happy."

"Except you?" she tried.

He put his hands up. "Hey, I trained to work with animals and I'm still working with one. I'm lucky.

When Anita goes, they'll probably stick me with the last captive mongoose or put me in the gift shop."

"So, good for everybody," she said. She looked skeptical, a single eyebrow raised.

He could tell she wanted to have one of those conversations he hated. Wasn't it so tragic? Didn't he feel terrible all the time? Didn't he want to let the tiger go, into the wild? To this last he always wanted to respond: even if that weren't the most asinine idea I'd heard a thousand times, where would this "wild" be?

He opened his menu and scanned the first page. "Good for everybody."

"Except the tigers."

He looked up, his pulse quickened. Why had he agreed to this? "Now see," he said, "that's just being sentimental. What tigers?"

"You're not serious."

"I see it all the time," he said. "Twenty times a day, the tears. So much easier to cry over a single tiger than the world they used to live in."

"What's wrong with that?"

"It's sentimental."

She gave him a look of bewildered irritation—eyes widened, brow furrowed, slight shake of the head—as if to say: can you really be this insensitive? "But isn't she lonely?"

He sat back in his chair. The woman didn't get it, and she wasn't going to get it. "Anita's fine. She doesn't know what she's missing."

"I don't believe that."

He signaled the waitress. "Look," he said, "she's not the first creature to outlive her kind. They all went the same way. One left, then none."

The waitress approached the table and asked what they'd like.

"He might be ready," the woman said, pointing at him without looking at him, "but I'm not."

"Beg your pardon," the man said to his date.

The waitress smiled. "I'll give you a few minutes."

She left them alone at their table, and the woman opened the elaborate menu, taking her time, reading up and down the cursive script in each column.

If she doesn't wipe that look off her face in thirty seconds, he told himself, I'm standing up and walking out.

She closed the menu and leaned back in her chair and crossed her arms. "I'm ready."

"No hurry," he said.

"They're not cheap, these old-timey menus."

He couldn't disagree. "These ancient traditions," he said and smiled and tapped his satin-backed *carte du jour.* "They have their price."

"Does it feel like, a little indulgent?"

"Do you want to go somewhere else?" His shoulders tensed and he looked at her intently. That would just drag it all out.

"No, no," she said, and raised her manicured hand.

"Good," he said.

The waitress returned to take their orders.

"How is this pepper jelly salad?" The woman asked.

"House favorite."

"Real cucumbers?"

"Absolutely real."

"Grown in dirt?" the man asked.

The waitress smiled and shook her head. Very funny. "How spicy do you want it?" she asked the woman.

"Oh, oh no. Is it spicy?" She looked from the waitress to the man and back again. "Can you make it like, not at all spicy?"

The waitress smiled blankly. "I'll make a note." She took the man's order and collected their menus. "Can I get you another drink?" She asked the woman.

The woman looked at her glass and glanced at the man. "I'm ok for now."

"I'll take another," the man said.

"You know," the woman lifted her glass. "I think I will have one more."

The waitress left them alone, and they exchanged another polite smile.

"You know the spice is really what makes that dish," he said. "Right?"

She put her hand on her belly and made a face. "I prefer it mild."

"The spiciness is the whole point of the fresh cucumbers."

"I prefer it mild."

"I've never heard of such thing."

"Well. Now you have."

"Now I have."

He could see very clearly that she was an idiot, strung way too tight. And she knew him for one of those cold-hearted sons of bitches, incapable of empathy, so sure of himself. What a disappointment. The man who works with the world's last tiger. He might have been a catch.

"You know, it'd be one thing," the woman suddenly said, "if these animals were just extinct." He closed his eyes a full second and inhaled through his nostrils. "I mean like, I shed no tears over lost dinosaurs. But these animals are gone because of us. It's our fault. Don't you consider that?"

The man shook his head. When she said our fault, something in her tone was saying his fault. He'd gotten this from people before. Anyway there was a thinking error in the whole idea—a lot of superstitious guilt. He told the woman about the tools they used to communicate with nonhuman animals—surely she'd heard about it?—and the message the last elephant had relayed from captivity in Germany. It was binary language that corresponded more to Spanish, say, than to English.

"So what does that mean?" she asked, shaking her head with her brows and forehead scrunched up.

"For example," he said, "in Spanish you don't say: I dropped the cup. You say, the cup, it fell from me. Se me cayó la copa."

She looked at him stupidly.

"The agency of the I is minimized," he explained. "It's different."

"I don't get it."

"'There is sorriness.' That's how the elephant's emotional state was translated. It didn't say 'I'm sorry.' Or 'You should be sorry.' It didn't place any blame."

"I don't understand," she said. "Anyway I don't agree."

He nodded. He could see that. She didn't agree with the elephant. Good for her. She went on, insisting: humans were the problem; it was wretched being human, and here they were: human. She laid her hands flat on the round table.

"I enjoy being human," he said. "I enjoyed my drink, I'm enjoying the music, and I'm enjoying the prospect of a good meal." Everyone around them seemed to be enjoying their humanity as well. Behind her, where she couldn't see, a couple had their hands clasped together over their own tiny table. He imagined that man's knee between his date's, beneath the table.

The waitress set down their new drinks and took the man's empty.

"I'm just saying, we are the problem. In this particular case."

He picked up his new glass. "I get so sick of this kind of talk, you know?"

She agreed. There were two kinds of people: kind, compassionate human beings, and jerks. Conversation across the divide was impossible and pointless. Always had been. This was why they'd come to this in the first place. "Never goes anywhere," she said.

"Never does."

They sat in silence a full minute.

"You said on the phone last week that you live just north of here?" he asked.

"The Rancho."

"Nice," he said.

"You?"

"What? Oh," he said. "Downtown. Right around the corner."

"Oh," she said. "Is that why you picked this place?"

"I'm sorry?"

"It's just that I'm going to have a long commute home."

For another full minute, silence. Two minutes. Laughter, music, and crystal rang around them. She could feel the bass in the music behind her sternum—she wished they'd turn down the volume on the trio—and the drinks were warming her face. He was leaned back in his chair, away from her. He wished they'd program a drummer into the music. It'd been nagging him, what was missing. Across the table the woman's eyes had glazed over, her elbow propped in her opposite hand, the ice in her drink catching the light.

"Only three people in the world have keys to her space," he said.

"The tiger?"

"Anita, yeah. Me, the so-called curator of mammals, my boss, and the head tech designer."

"The guy who makes all the exhibits?"

"The interactives, yeah. And he's an idiot."

"My boss is an idiot too."

They raised their drinks to that and laughed.

"Well this guy isn't my boss, exactly. He's all of twenty-five years old," the man said. "He comes in and designs the forest with rivers and mountains, like he was hired to do, and then he populates them with tigers using Anita's data. Figures out cubs with a combination of cat and puppy data—regular old golden retriever—and puts it together with Anita's. But on the sly, he makes an interactive exhibit with hoops of fire and leaping tigers and all that. It's in a cityscape, with schools, churches, houses. It's ridiculous. City or circus? They didn't even know what to call it."

"Hey, I've seen that billboard."

"Yep. But he didn't tell anyone he was doing it at first. He's just a smart-ass kid, messing around with his software, right? Curator of mammals goes ballistic. Threatens to fire the kid, with intent to sue on some kind of software-use issue."

"Yikes."

"Then we find out over the course of the day that the city circus outsold the boreal forest by 800 percent. And by 2500 percent the following day."

"Hoops of fire."

"Right. Turns out people like them. Little boys and college kids love it."

"Have you ever been hurt?"

"By Anita?" He lifted his right arm. "Broke my arm, oh, four years ago. It was my fault, and it was mostly the horse pelvis that did it."

"Horse pelvis?"

"Tiger snack. Really fucking heavy. Excuse my French. Every now and then Anita will still get a little sketchy around food."

"But otherwise she's gentle?"

"Well it's not like she naps with her head in my lap or anything like that."

The waitress brought their plates.

"Listen," she said, lifting her fork. "When we finish, why don't you take me to see her?"

"Anita? Tonight?"

"Yeah."

He laughed, picked up his fork. "I don't know. She's old, and right now," he looked at his watch, "she's probably sleeping, hiding somewhere. Besides. Zoo's closed."

"Oh come on."

"Anyway," he said. "I never actually show women the tiger until after the third or fourth date. So I know they're committed."

"To seeing the tiger?"

"Right."

They ate their food. The waitress refilled their water glasses and he asked her about her own job, which she told him she found dull and temporary.

"So," she said.

"So," he said.

The waitress took their plates and returned with the bill. He took it. "Sorry about the long commute home," he said. He really was. It hadn't been intentional.

She shrugged. "It's all right. Thanks for the dinner." She smiled. "It was good mild like that."

"I'll have to remember that."

"I'll think about the things you said."

He nodded. So ok, they were being kind now that they were saying their goodbyes. "Tell you what," he said. "Come by in the morning."

"Really?"

"Come right at eight before the kindergarteners start lining up and I'll introduce you."

"And I can touch her?"

"Sure, if she lets you."

▼

The following morning was clear, but Anita wasn't visible in her two acres of water, trees, and ledges. The rocks and boulders were flecked with mineral glitter, harvesting sunlight to run the interactive exhibits all the children would be playing in that morning. The trees were a pretty blend of wood and tech-cells to absorb the dawn moisture—and rainwater, if there was any. The single mat of hearty weeds—cheat grass and kudzu—was absolutely wild, an indulgence in terms of water use but which they allowed because it had always been where Anita chose to sleep.

The woman was late. It was twenty minutes past eight and the kids, their teachers, and some member of tech support were due along any moment. Finally, in the distance, on the path that came up over the hill from the zoo entrance, he could see her approaching in a yellow blouse and blue jeans. Practically alongside her, a few feet back, a rope of small children holding hands led by tech support and their kindergarten teacher.

Now Anita was on her feet, out from behind a boulder, her nose and eyes lifted in the direction of the crowd.

"Oh," the woman said, rushing up toward the man, but her gaze already fixed on the tiger. "There she is."

"Morning," the man said. "Welcome to the land of shitting, breathing tiger."

She laughed. "Good morning!" She flashed a brilliant smile.

Anita yawned.

"My God, her teeth," the woman said.

"You have no idea how much care those teeth have had," the man said. "Those two canines alone, tell you what. More dental care than everybody else in the city combined."

"I'm sorry I'm late," she said.

He shrugged and gestured toward the kids. "If you can stick around till they go, I'll take you in."

The line of children broke up as they let go of each other's hands and crowded around the gate. The man and woman kept off to the side.

Tech Support this morning was tall and lean with balding head and halo of red hair. He nodded a friendly greeting in their direction, and began his memorized speech about gathering data from live animals for the software in his office.

"How old is she?" the woman whispered to the man.

"Sixteen."

"Is that old?"

"Older than she would've been without us," he said. "She's slowed down some. Used to find her in those trees every morning."

"That must have been something to see."

"She'll be with us a couple more years."

"Then what?"

"We have all her data."

"And they'll make more tigers with it?"

"Yeah," he said, and laughed. "That whole cloning thing is not exactly working out."

The tiger began to move slowly toward the fence, stretching out one back leg at a time as she moved.

"So beautiful."

"Yes."

"When do we get to see the real tigers?" A child asked in a shrill voice.

"These are the real tigers, stupid," another boy answered for him.

"There's only one real tiger," Tech Support corrected, positioning himself before the children to explain again how the zoo worked. The teacher shushed them. "This is Anita, who shares all her information about being a tiger with me, so that I can put it in my computer, and create the tigers you're all going to see out on in the northern boreal forest, in a few minutes."

"Don't we get to see the circus?"

The kindgergarten teacher smiled. "When you come back with your parents. Today's a schoolday. We're going to learn about the forest ecosystem."

"Do we get to shoot them?" A boy asked.

"There's no hunting at this zoo," Tech Support told him.

"My cousin lives in Ohio and he gets to shoot them."

"We don't do that here."

The tiger closed her eyes and rolled on her side.

"Can we put our hands in their mouths?"

"You can try," he said, "but even our virtual tigers won't let you do that." He nodded at another raised hand.

"Does he do anything?"

"She. And yes, she does. She is a real live tiger."

"But what does she do?"

Another hand. "Would she kill us if we went in her cage?"

"She's a very powerful animal."

"Could she bite off our heads?"

"We'll have time for more questions as we walk," Tech Support said, turning his long lean frame to the trailhead. The children and kindergarten teacher arranged themselves in some pre-determined order, joined hands, and in a string of bright jackets and t-shirts filed behind him, shuffling toward the first of the interactives: Reptiles.

The man turned to the woman. Her dark hair fell in curls around her shoulders. "You want to go in?"

"Is it really ok?"

"Come on."

"Will we feed her too?"

"No. Feeding and new company isn't a good idea." He opened the gate, then looked back at the woman. She shrank back, and he took her elbow to encourage her. They stepped inside. He closed the gate behind them, led her a few feet in, then stooped down to the ground and waited for the tiger to approach. "Come on," he said, waving her down. "It's ok."

Anita came straight to the man and rubbed her face against his knee, and lifted a paw onto his thigh. He steadied himself as he talked to the tiger. Beside him the woman held her breath high in her chest and looked from the man to the animal with wide, nervous eyes. He took her hand, and ran it softly along the side of Anita's face. Anita turned toward the woman, and suddenly dropped low, her rear end high, her tail wagging.

"She's saying hello. Go ahead. Touch her face."

When the woman put her hand on the tiger, she took a deep jagged inhale and her eyes glistened. Her free hand fluttered to the top of her chest, then over her mouth. "I didn't expect," she said, glancing at him. "My God."

"Wish you could have seen her in the trees."

"She is so enormous. She's magnificent."

"That's our Anita."

Anita dropped to the ground and rolled over on her side. The woman raised both hands to her mouth and shook her head. "I don't know how you can bear it, seeing her everyday." Then she lowered her voice. "I feel like she can hear us."

"Of course she can hear us." He laughed. "She's a tiger. Not a deaf-mute."

"Doesn't it just break your heart?"

He shrugged. "No." Anita was on her feet again, sniffing the air.

The woman's face crumpled, her nose and mouth twisted up and she paused, bit her lip. "Isn't it just too sad?" The tiger was circling away from them, toward the water.

He smiled and shook his head at the woman.

"I'm sorry," she said, "I'm sorry I'm being silly. You hate this kind of talk."

"It's alright. Don't hold back on account of me."

"Ah, God. Look. She's looking at me," the woman whispered.

"She'll do that." Beyond the wall they could hear the faraway chatter of another group approaching. "So, I'm sorry but they shouldn't see you in here." He nodded in the direction of the sound.

"They'll want to come in too?"

"That and I'd get fired."

▼

"I hope you don't think I used you to see the tiger," she said, back outside the outermost gate. They stood with their hands at their sides.

"Nah." He said. "Zoo's open to the public. Come on by any time."

"Well." She extended a hand, and they shook. "It was so good to meet you. Thank you. I'll never forget this."

"Like I said. Come by and see her any time."

▼

When the woman was gone, he looked out into the acreage where Anita had stretched out beside the water. Even just a few years ago, she'd have spent all morning up in the trees, or splashing and half rolling in her water pools. It'd been longer than that—over a decade—since she'd known another tiger, felt the companionship and warmth of another of her own kind. The man wrapped his fingers around one of the steel bars. It was smooth and cool. He looked in at Anita, a little tiredness burning the backs of his eyes. "That's my girl," he said.

CODA: LOVE

"Love…bears all things, believes all things, hopes all things, endures all things." These words, written by the Christian apostle Paul to the church in Corinth almost two thousand years ago, are about as familiar a sentiment regarding love as we are likely to get, rivaled only by Virgil's "love conquers all," and the Beatles' "all you need is love." Such aphorisms are the stuff of wedding vows, greeting cards, and adolescent poetry, yet complexities lurk behind the simple words.

▼

Among the several Greek terms that are sometimes translated by the English word "love," Paul uses "agape" which carries no suggestion in this passage of eroticism or of love's physicality. "Amor," the Latin word used by Virgil in his pastoral poem "The Eclogues," has broad meaning, but here refers to love as an irresistible erotic force. The Beatles' song, which debuted in 1967 to 400 million people in twenty-five countries in the first live, global television link, overflows with the universal love ethic of the 1960s counter culture. What we can see already is that love is both revealed and obscured by the simple language in which it is often expressed.

▼

Across cultures, languages and centuries, love has shown itself as a flux of shifting beliefs, feelings, ideas, actions, and cultural meanings rather than as a timeless concept with a universal essence. In its various forms and manifestations, it is the subject of centuries' worth of painting, music, and poetry, and some of the world's major religious traditions claim it as their focal point and common ground. It has inspired war, peace, civil and human rights movements, and is the subject of intellectual inquiries ranging from history, philosophy, and sociology to psychology, neuroscience, and biology. Love takes diverse objects including friends, parents, partners, pets, children, places, nature, and countries. Most of us care deeply about having love, losing it, getting more of it, wondering whether we give enough of it, struggling to understand what it is, when it is healthy and appropriate, and on and on and on. For many of us, love is a central preoccupation of our lives. Everything else can seem a waste of time.

Most of us would say that love is constant, whatever else it is; fair weather love is no love at all. And we would insist that the beloved—whether partner, parent, child, or pet—is irreplaceable. We may come to love a second partner, child or pet, but these are distinct loves, each with their own story, not just another installment in our own domestic lives.

Hovering in the background behind these declarations of all-important, constant, irreplaceable love, is the often inchoate recognition that any particular love of ours is radically contingent, even though what love demands seems highly specific to the one we love. Your parents would have loved anyone who turned out to be their child—it didn't have to be you—yet it is you they love and sacrifice for, no one else. If a different person had responded to a man's personal ad with a wittier response or a more alluring photograph, he might now be living happily ever after with her. Yet the one who did respond is now the love of his life and the person he favors in all his actions, even if by the lights of the world others are more deserving.

▼

This man's ability to "love the one he's with" may be pointing to what Paul the Apostle and Paul the Beatle were hoping we'd all realize: the possibility of a transcendent love, not directed toward a particular object and subject to chance or contingency, but a big love that can be cultivated and practiced in the way you might exercise a muscle. One who aspires to love in this way works hard to develop the ability to bring love to all the creatures she encounters (regardless of who in particular answers her personal ad). This kind of love would place no limit on the number of people, places, animals, or things that can be loved.

▼

Whatever our aspirations, the sad truth is that we betray our lovers, abuse our children and parents, and treat our animal companions as if they were extensions of ourselves with no needs of their own—not always, but often enough. Almost any idea of what love is can be easy to accept in theory, but in the conflicts and contradictions of daily life it is not easy to embody love in practice. As so many of our relationships have taught us, loving is a resolve we must return to again and again—a hundred, even a thousand times a day—and almost never is it the simple, head-over-heels, heart-stopping, once-and-for-all experience we may fantasize about. In the gritty and difficult days and nights filled with the people we actually meet rather than those we might have met, of the country or institutions into which we were born rather than those of distant strangers or enemies, love may seem elusive—in its meaning, in the extent to which we are capable of it, and in the degree to which it is contingent upon history, culture, time, and place. Sometimes it seems that there is something terribly unsteady about love as we make and experience it—something so fluid and variable, and so difficult to keep at the center of our lives.

What makes loving so hard to understand and even harder to practice? The novelist-philosopher Iris Murdoch points to an answer when she writes, "Love is the extremely difficult realization that something other than oneself is real."

▼

This realization that Murdoch calls "difficult" may seem obvious but it is not. Love is the antidote to an all too familiar narcissism—one that mutates out of a necessary and healthy self-love—that is a master of disguising itself just where we think love might be thriving. A high degree of self-absorption may have enabled human beings to survive in primal struggles for food and shelter, and may still help many of us get through days and nights that would otherwise seem crushing or demeaning. This kind of self-centeredness may have its place and purpose, but the job description of an overbearing, even tyrannical ego that may serve human beings well enough in difficult times is in need of demotion from CEO to Occasional Advisor if we are ever to achieve the best of times. A ravenous ego can devour the very self whose interests it is supposed to serve. Who hasn't sensed this in the midst of a challenging relationship? Even when we're in love (or so we think) we find the dear self intruding like a jealous partner, trying to sabotage the possibility of intimacy with another. We see ourselves in others, but all too often see only ourselves.

▼

Murdoch's torturous realization, when it occurs, can come in extremely different forms: the realization that one's puppy is real; that one's lover is a distinct person; that the reality of the scorching winds of the desert cannot be denied; that an artist can speak to us across the centuries in her own pitch, timbre, and voice in a way that can change our lives. Murdoch's realization can even come when we pass strangers on the street, look into their eyes, and recognize that they too have experienced love, loss, fear and joy as poignant and crushing as our own, though we may never know their stories. In addition to recognizing the fullness of others, love also involves sustaining sometimes uncomfortable, challenging, and fiercely intimate relationships. The paradox of love emerges in its demand to see the beloved as independently real, yet part of a larger conception of oneself.

Once we grasp Murdoch's point we see the most profound challenge of love in the Anthropocene. The Anthropocene threatens to give us a narcissist's playground—a nature that is only the extension of ourselves and our desires, without independent meaning or sustenance. Loving relationships are not possible in a world that consists only of oneself and one's projections.

Think about that old song, "To Know Him Is to Love Him," first recorded by The Teddy Bears in 1958. On first hearing, the song can seem insipid, but as the music critic Robert Chistgau points out, when you listen to the 2007 performance by Amy Winehouse, you feel that a profound truth is being revealed: to know him is to love him, and to know the song—really know the song—is to love it as well. And what is true of songs is of course true of mountains, deserts and nature itself, as well as the people and animals with whom we share the Earth. To know them is to love them. But not everyone who experiences them will know them. And not everyone who knows them will really love them.

There is an intellectual kind of knowing we're all familiar with: "Oh sure, I know the Snake River. Wyoming, right?" And there is a deep knowing that entails bodily experience and intimacy: "I know that river—every bend, every shallow; it's where my partner and I honeymooned, where my daughter caught her first fish, where I swim naked as the rushing snowmelt levels off every late summer." This kind of knowing can sometimes happen with a single life-changing experience, but more often it takes continual resolve over time, and requires us to look again and again even when we think we already know what we're seeing. Everyday at 7 pm your partner walks through the door after her work day; what greater kindness, what greater act of love than to greet her with a mind open enough to ask yourself:

and who is this? But even this kind of experiential knowing may not result in love. That daughter who is brought to the Snake River every August with her mother and father may not love it at all, though she knows it very well. She might appreciate it, she may have experienced it as her parents have, but if given the chance, she'd rather be skiing, or biking in the city, or decorating cakes. When it comes to love, much is possible, but nothing is certain.

▼

In many ways the task before us in the Anthropocene is the same task that has always been before us: to get the dear self out of the way enough to be able to really see and come to know—in relationship— the world of other people, plants, animals, oceans and rivers around us. But there may be more at stake, now, as well.

In the world as we have known it, nature is a partner even in many of the loving relationships we have with each other. Remember Make-Out Point, on a cliff looking out over the sea. We take our children camping in parks or wilderness areas. Families have cookouts by the river. Kids explore the magic forest just beyond where the backyard ends. In some cases, nature is explicitly part of the object of love: surfing, climbing trees, swimming in lakes, walking in the woods, scrambling up mountain peaks are all unthinkable without nature. And sometimes nature itself is the beloved: the Bitterroot Mountains, the Sahara sands, the crystalline waters of Five Flower Lake in China, all have their lovers.

▼

As the Anthropocene becomes more fully present, as the world and technology change, some familiar human experiences of love and the natural world, even modest ones, may become increasingly unthinkable, even lost. Consider John Sebastian's 1966 song "Rain on the Roof," which describes falling in love with someone "underneath a roof of tin" while the rain "soaks the flowers." Imagine a world of endless drought: no rain, no tin roofs, no soaked flowers, and a lot more from our familiar world missing besides. Will experiences like this or the art that expresses them be accessible to us? Will we even be able to understand the loves of our parents or grandparents?

The general idea that time and circumstance may make particular expressions of love seem distant and foreign to us is familiar. The passionate love that so many heroines of eighteenth and nineteenth century novels felt for their cousins is recognizable to us, but can seem as odd and strange as the love that some report in arranged marriages. Some people have difficulty seeing love in relationships that have at their center sexual acts that they consider degrading, perverse, or immoral.

▼

But love in the Anthropocene will present challenges that are even more extreme, and which may make the ever difficult experience of loving even more disorienting and elusive than ever. Think, for example, of the movie *Her* in which Theodore Twombly (Joaquin Phoenix) falls in love with an operating system (Scarlett Johansson). Or if this seems far-fetched, reach into your real or imagined memory and recover those feelings of joy and wonder as you were careering down double black diamond runs in Tahoe or Vail. It was just you and the mountain, you and the wind; no condos and no snow-making machines. Now imagine that Monsanto has brought you "Ice-9," a nanoparticle that forms the nucleus for fluffy snowflakes that don't melt and stick. The snow-making machines are gone, and every day is a powder day. Is this better? Do you love the experience more? Or has the object of your love slipped away?

We may once have thought of nature as the backdrop against which we lived our lives—

pretty scenery to be sometimes plundered for its resources, or to write songs about—but the Anthropocene throws into stark relief that nature is not a "background" and never has been. It has always been a part of our lives—a part of us as we are part of it. In the Anthropocene, the world is increasingly one of our own making—one that we do not know how to see, know or love, in Murdoch's crucial sense—not even when we have the best of intentions. A natural resource manager may restore what he thinks of as a wild landscape, but this "wild" landscape answers to our fantasy of the wild, and is wholly dependent on human ingenuity and intervention for its very existence. In losing nature as a fully independent partner, we will have lost our best teacher for learning that something other than ourselves is real and thus a profound opportunity for learning how to love.

As the force of the Anthropocene is increasingly felt in our lives, important questions about sustainability, environmental quality, and the preservation of nature and heritage become ever more urgent. The languages of science, technology, and economics dominate the discourse: can it be done? How much will it cost? But there are other questions that the Anthropocene will ask and other languages in which they must be discussed: should it be done? What will we become if we follow that path? The Anthropocene will challenge not just our science and technology, but also the human heart in ways that are difficult to predict but which we're already beginning to experience. The question we ask may seem simple but is fundamental: how will love arise in a world without nature as we have known it?

ACKNOWLEDGEMENTS

We would like to thank Kate Johnson and John Oakes for believing in this project; the School of Social Science at the Institute for Advance Study in Princeton for the fellowship that made Dale's work on this book possible; Leif Wenar, Katie Holton, and Dillon Cohen for organizing readings; and Andrew Chignell for reading early drafts.

ABOUT THE AUTHORS

Dale Jamieson has held visiting appointments at the National Center for Atmospheric Research in Boulder and the Institute for Advanced Study in Princeton. He is currently Professor of Environmental Studies and Philosophy, Affiliated Professor of Law, Affiliated Professor of Medical Ethics, and Director of the Animal Studies Initiative at New York University. He has published widely in environmental philosophy, animal studies, and ethics: most recently *Reason in a Dark Time: Why the Struggle to Stop Climate Change Failed—and What It Means For Our Future* (Oxford, 2014). This is his first work of fiction.

Bonnie Nadzam has published fiction and essays in many journals and magazines, including *Granta*, *Harper's Magazine*, *Orion Magazine*, *The Iowa Review*, *Epoch*, *The Kenyon Review*, and many others. Her first novel, *Lamb*, was recipient of the Center for Fiction's first novel award in 2011, and was longlisted for the Baileys Women's Prize for Fiction. It has been translated into several languages. Her second novel, *Lions*, will be out from Grove Press in 2016.

Also from OR Books

Welcome to the Greenhouse
New Science Fiction on Climate Change
EDITED BY GORDON VAN GELDER

"No matter what you believe about climate change, *Welcome to the Greenhouse* is a treat for anyone who appreciates good SF in the best speculative tradition." —ANALOG SCIENCE FICTION AND FACT

Ivyland: A Novel
MILES KLEE

"*Ivyland* is a harsh, spastic novel about drug-addled misfits clawing their way through a wrecked future that feels disconcertingly familiar. As if that wasn't enough, it's also got evil caterpillars, flung jellyfish, great prose, new drugs, sharp jokes, a stolen ice cream truck and a miracle tree." —JUSTIN TAYLOR

Watchlist: 32 Short Stories by Persons of Interest
EDITED BY BRYAN HURT

"A brave and necessary set of early flares of the literary imagination into the Panopticon we all find ourselves living inside these days." —JONATHAN LETHEM

www.orbooks.com